SAS Publishing

P9-APN-184

Web Development with SAS® by Example

Second Edition

Frederick E. Pratter

THE
POWER
TO KNOW.

Contents

Preface

SAS Web Technologies provide a comprehensive set of tools for creating and deploying Web applications. This book is an attempt to bring together in one volume a set of examples to illustrate the major issues in Web development, using the tools available from SAS as the building blocks. It is intended both as a general introduction to Web programming and a guide to using SAS Web tools. As such, it is aimed at least three different audiences.

The first group includes experienced SAS users who want to get started delivering content on the Web. These are the attendees at SUGI and the regional conferences who fill lecture rooms and workshops for sessions with SAS/IntrNet, SAS Integration Technologies or SAS AppDev Studio in the title. The second, complementary group includes of Web developers who are interested in access to SAS data and who appreciate the ease and simplicity of the solutions SAS has provided.

The third group includes those project managers, students, and others who want to learn about Web development and who are not well served by the current proliferation of specialized texts. Clearly it is not possible to cover this entire topic comprehensively in a reasonably sized volume. On the other hand, introductory users are often baffled by the wide variety of options available and the blizzard of confusing Web terminology.

This book, then, attempts to achieve two goals: first, to organize the options for Web programming into an understandable framework, and second, to show how the available SAS tools fit into this framework. The strength of SAS is its ability to make difficult data analysis and presentation tasks straightforward, if not simple. With SAS Web Technologies, users can leverage their substantive experience without having to learn a whole new specialized set of tools. In addition, the full power of SAS is available for data exploration, analysis, and reporting.

A Note to the Reader

A new version of SAS AppDev Studio (3.2) is planned for release later in 2006. The webAF and webEIS components will be going away in the new version of SAS AppDev Studio; instead SAS is moving to *Eclipse*, an industry standard, open source development platform (see http://www.eclipse.org/ for more information). Since many of the examples in this volume use webAF, initially SAS Press considered delaying the new edition a year or so to update it with the new version of SAS AppDev Studio. Finally, however, the benefits to users from getting the book out as scheduled outweighed the disadvantages, so it was decided to proceed as planned. The material covered in this volume is up-to-date for SAS 9.1.3 and SAS AppDev Studio 3.1.4, and it should serve as a comprehensive introduction to SAS Web Technologies as of the date of publication. The good news is that plans for the third edition are in the works.

What Is Covered in This Book: Changes in the 2nd Edition

Every example in the book has been reviewed to make sure it works with the most recent versions of the software. All of the material from the first edition is still included, though discussion of some of the deprecated products has been moved to the appendixes. In addition, a substantial amount of new material has been added—in particular, an entirely new section on features introduced or enhanced in SAS 9.1.

In general, this book covers four main topics:

- getting started with Web programming, including using the Output Delivery System to create static HTML content
- programming with SAS/IntrNet using the Application Dispatcher and htmSQL
- programming with Java and SAS AppDev Studio
- SAS Integration Technologies and the SAS Open Metadata Architecture

There are specific SAS components for each of these functions; each is described in some detail in the chapters that follow. The focus, however, is not on the specific products, but rather on examples of commonly (or uncommonly) performed Web development tasks and how these are supported by SAS.

The approach throughout is progressive. Part 1 begins with a general introduction to Web programming and the Internet protocols. The remaining chapters of the first part include a simple HTML tutorial and a description of SAS support for static HTML pages using ODS, the SAS Output Delivery System. Readers who are already familiar with Web programming in general and HTML can skip straight to the discussion of ODS in Chapter 3, "Creating Static HTML Output."

Part 2 covers server-side applications programming using the Common Gateway Interface (CGI) and Java. The first chapter in this part is a discussion of how to connect to a remote SAS data server using the SAS/SHARE and SAS/CONNECT components. Many of the subsequent examples require a connection to a SAS server; this chapter has been included as an aid for Web developers who are not familiar with these products. In addition, the SAS/CONNECT driver for Java provides the Java classes that communicate with a SAS/CONNECT server, while the SAS/SHARE driver for JDBC enables you to access and update SAS data through Java programs that access a SAS/SHARE server.

Chapter 5, "Web Applications Programming," includes some simple Perl programming examples and may be useful to those who are interested in learning about CGI in general. The remainder of this part covers the two main technologies included in the SAS/IntrNet product: the Application Dispatcher and htmSQL.

Part 3 explains how to use Java to create dynamic applications using Java servlets, JavaServer Pages (JSP) and custom tag libraries. The focus is primarily on using webAF software to create JSP.

Part 4 has been added to introduce the new SAS Open Metadata Architecture, the Integrated Object Model (IOM), and the use of SAS Stored Processes to build Web applications. The new SAS BI component, SAS Enterprise Guide, can be used to create and test stored processes.

Part 5, the appendixes, covers a variety of topics that in general represent older approaches that either have been superseded by the methods described in the body of this book, or else (as is the case with the Design-Time Controls) have specific, narrowly defined functions. These topics include the following:

- client-side dynamic Web content with DHTML and JavaScript
- Java Applets
- SAS Design-Time Controls, used with third-party HTML editors to provide SAS connectivity
- webEIS software for generating OLAP reports

As should be obvious from this brief introduction, SAS Web Technologies provide an extremely powerful and flexible collection of tools for Web development. There are also a lot of moving parts, and new functionality is being added on an ongoing basis. The goal of this volume is to try to put all of the pieces together in a systematic overview, so that both novice and experienced Web programmers can find the information and examples they need to get started using these tools effectively.

What is Not Covered in This Book

In addition to the products previously described, a number of important Web technologies not covered in this book are included with the SAS Enterprise BI Server suite (see http://support.sas.com/news/feature/04dec/biserver.html for a more information about these components).

Documentation for these components is available on the SAS Web site:[1]

- SAS Add-In for Microsoft Office – provides to access to SAS directly from Microsoft Office via integrated menus and toolbars (http://www.sas.com/technologies/bi/query_reporting/addin/)
- SAS Web Report Studio – provides query and reporting capabilities on the Web (http://www.sas.com/technologies/bi/query_reporting/webreportstudio/)
- SAS Information Delivery Portal – provides a single access point for aggregated information (http://www.sas.com/technologies/bi/content_delivery/portal/)
- SAS Information Map Studio – creates information maps to expedite access to organizational data (http://www.sas.com/technologies/bi/query_reporting/infomapstudio/)
- The SAS Web OLAP Viewer for Java – provides a Web interface for viewing and exploring OLAP data (http://www.sas.com/technologies/bi/query_reporting/webolapviewer/)

The SAS Web Report Viewer and the Visual Data Explorer are client tools that are available separately, but which require the BI Architecture on the server. SAS Press currently has a "BI Cookbook" under development that will cover the BI components in detail; this book focuses on the tools included with SAS AppDev Studio, SAS/IntrNet software, and SAS Integration Technologies, along with Base SAS software components such as ODS and the HTML formatting macros.

A Note on the Examples

All of the examples in this book were created using resources that are generally available. The data sets SASHELP.SHOES, SASHELP.RETAIL and SASHELP.PRDSALES, used for many of the sample programs, are all included with the basic SAS installation. This has two implications: first, the examples are not terribly exciting. Second, however, all of the code illustrated can be executed in your local environment. No special downloads or customization is necessary to run the sample programs, assuming that the necessary products are licensed at your site. It is hoped, thereby, that you will be prompted to try out these examples, and use them as the basis for more sophisticated applications.

---

[1] Since the SAS Technologies and Support sites are frequently updated to reflect new products and solutions, these references may change in the future, but they are correct for the date of publication.

The SAS AppDev Studio setup routine assumes that you will be using a Web server and SAS data server (such as SAS/SHARE or SAS IOM) on the same Windows PC as the Web development environment. This is fine for Web page development, but unrealistic for training, since most sites will probably have separate Web and SAS servers. As the first chapter of this book points out, a more representative test environment, and one that is not terribly difficult to implement, is to use two or three PCs: a Web server, a back-end data server, and a client system. A multiplatform setup more closely simulates a real client-server environment, since setting up connections to remote systems involves a great many issues which do not come up if the client, the Web server, and the data are all on the same PC.

In order to simulate the range of platforms for which SAS is available, the examples in this book use a small private LAN created specifically to test the programs shown. Several hosts are available:

- *Odin* – the client system, running Windows XP Professional SP2
- *Hunding* – a Windows 2000 server, with Microsoft Internet Information Server (IIS) 6.0 as the Web server
- *Hygelac* – a Red Hat Linux server platform for the Apache and Tomcat Web servers

SAS 9.1.3 SP3 was used on all of these; SAS AppDev Studio 3.1 is installed only on the Windows XP client *Odin*. In general, all of the examples presented were run on all of the platforms that were appropriate in order to increase the probability that they will work at your site.

Acknowledgments

For the second edition of this book I want again to thank John West, my editor at SAS Press, for his superhuman patience as I struggled to figure out how all this stuff works. Thanks also to the folks at SAS Press for their care and dedication in shepherding the book through the publication process, not to mention some splendid dinners.

I owe a particular debt of gratitude to the technical reviewers who patiently corrected and sometimes re-corrected my errors in the manuscript. In particular, Vince DelGobbo helped me to understand proven technologies like SAS/IntrNet and the Design-Time Controls, as well as the new SAS Stored Process Web Application. Mickey Miller answered countless questions about using Java with SAS. Joy Lashley reviewed the entire manuscript with incredible attention to detail and certainly saved me from many embarrassing blunders. Rich Main was also helpful in developing my understanding of the strategy and utility of the products under his direction. Additional SAS technical reviewers I'd like to thank include Tammy Gagliano, Scott Sweetland, Anita Hillhouse, Cynthia Zender, Bari Lawhorn, and Chevell Parker. Whatever errors or infelicities that remain are mine, but the resulting product is far superior to what it might have been without their help.

I also appreciate the support I have received from my friends in the SAS User community. In particular, Elizabeth Axelrod and Ray Pass at NESUG, Pete Lund and Curtis Mack at PNWSUG, Diana Suhr at WUSS, and Margaret Hung and Sandra Minjoe at PharmaSUG all provided sponsorship and encouragement as I presented various drafts of this material at SUGI and the regional conferences. Thanks also to Art Carpenter, the other *éminence grise*, for his friendship, sense of humor, and especially for his patience with my stories.

The faculty here at Eastern Oregon University deserve my gratitude for both moral and financial support. This is a wonderful place to work and think. Initial funding for research on the new technologies in SAS was provided by a summer stipend grant from the EOU Faculty Research and Grants Committee. Dr. Marilyn Levine, Dean of the College of Arts and Sciences, provides enthusiastic support for my research activities along with approving all of my conference travel requests without a murmur. Dr. Richard Croft and John Clever, my colleagues in the Computer Science and Multimedia Studies Program, covered my classes and made it possible for me to have some time to work over the two-year gestation process of this book.

Finally, to Linda, for her wisdom, patience and understanding. Thanks for sharing me with this book project.

Part 1

Getting Started with Web Programming

Chapter 1

SAS and the Internet

Introduction

SAS provides a powerful and sophisticated suite of products for Web application development. As is often the case with SAS, there are usually at least three different ways to accomplish the same task with the available Web programming tools. What is more, many of these tools are new even to experienced SAS users. In order to understand how to use these tools and what they do, the user also needs to be familiar with several other SAS products, including the Output Delivery System (ODS), Remote Computing and Remote Data Services, and SAS Integration Technologies. In addition, although it is not essential, it is extremely helpful to have some familiarity with HTML coding, CGI scripts, and Java programming in order to understand how the various SAS components work.

While it may be true that a wrench and a pair of pliers do pretty much the same thing, there are times when one will work and the other won't. Learning to use the tools that SAS has provided is largely a question of figuring out when the wrench won't fit. The goal of this book, therefore, is to introduce the entire SAS Web development tool kit, to explain what each component does, and to suggest when to use specific features and functions. Some familiarity with SAS syntax is assumed, specifically the DATA and PROC steps, but the discussion of Web programming begins with the most basic kinds of information.

Currently there are hundreds if not thousands of books about Web application development, ranging in coverage from the fundamental to the monumental; several of the more useful ones are referenced at the end of each chapter. Nonetheless, it is not easy to find a one-volume introduction to the subject of Web development that manages to combine comprehensiveness with intelligibility.

At the other end of the spectrum, it would be possible to write entire volumes about each of the topics covered in this book. Consequently, compromises had to be made about how much or how little detail needed to be included. The goal of this book is to discuss the design challenges and available solutions, and to attempt to demonstrate by example how the tools available from SAS fit into this conceptual framework. Note too that this is *not* a book about Web usability.[1] Content and page design are assumed. The focus in this book is on how to get the page designers' brainstorms to work in practice.

As used in the SAS documentation and in the rest of this book, SAS AppDev Studio includes the following product components: SAS Integration Technologies, SAS/IntrNet, SAS/CONNECT, SAS/SHARE, webAF, and webEIS. Since these products are not household words (at least not in most households), all of these additional components will be described in their place, as part of the overall conceptual framework that is the SAS Web development suite.

Strictly speaking, however, the new content in SAS AppDev Studio is just the two modules webAF and webEIS; both of these are available only for the Microsoft Windows environment.

- webAF is an *Interactive Development Environment* (IDE) for building Java applications to access your SAS data and present it on the Web.

- webEIS is an *Online Analytical Processing* (OLAP) report builder for the Web; it is a point-and-click application builder for creating documents and publishing them on the Web as Java applets or JavaServer Pages. (Note that webEIS has been deprecated in SAS®9, although it is still supported in SAS AppDev Studio 3.1.)

Don't worry if you don't know what an IDE or an OLAP is; we will get to them in due course. Learning about Web programming is largely a matter of learning to navigate through a haze of jargon. To explain what all these tools do, it is necessary to use a lot of TLAs (three-letter acronyms). There are also quite a few four-letter acronyms, and even some five-letter ones.

---

[1] A classic one-volume treatment of this topic is by Jakob Nielson, *Designing Web Usability: The Practice of Simplicity* (New Riders Press, 2000).

People have different styles of learning, but relatively few people are blessed with the ability to look at a page of documentation and come away with a picture of what the software is supposed to do. Consequently, much of this book consists of examples. The hope is that if you can decode what the documentation is talking about, it will begin to be useful to you, and you can proceed beyond the simple problems in this book. The remainder of this book takes up the challenge of defining these new technologies and illustrating how SAS tools can be used to create distributed information processing systems.

TCP/IP and the Internet

The *protocol* used to communicate among different computers is now almost universally the *Transmission Control Protocol/Internet Protocol* (TCP/IP). In general, a diplomatic protocol is a set of previously agreed-upon rules for negotiation. In order to send data from one computer to another, there must also be an agreed-upon set of rules for how that data should be addressed and formatted. Computers use different sets of protocols to manage this process. Each protocol is designed for a different purpose, depending on how much reliability and control is needed.

In the case of network-based data transmittal, the important concern is that *all* of the data arrive in the correct order. The TCP/IP protocol was developed back in the 1970s by a team of scientists working on the Department of Defense Arpanet (Advanced Research Projects Agency Network) project.[2] The original project had a number of goals, one of which was to assure that command and control messages could still go out and be received in the event of a thermonuclear attack on the United States.

In order to meet this requirement, the researchers created a protocol that would allow messages to be sent as discrete packets of information via any number of possible routes. They would then be reassembled at the receiving end in the correct order. The Internet Protocol, or *IP*, is responsible for forwarding the packets to the specified Internet address; *TCP* is the set of rules for sending and receiving packets over the physical network, and for catching and correcting transmittal errors. (See http://directory.google.com/Top/Computers/Internet/Protocols for a list of available resources on TCP and IP.)

The global network that became known as the Internet consists of a great many loosely connected clusters of networked computers, all using TCP/IP to communicate. The computers in your home or office network can all talk to one another using TCP/IP, and your network can talk to all the other networks in the world using the same mechanism. It should be noted that there are alternative networking protocols, most notably IPX, which runs on Novell networks, and LAN Manager, which was IBM's contribution. These alternative protocols are dwindling in use, however, due to the enormous impact of the Internet and the World Wide Web, which run on TCP/IP. In this book, the focus is on how SAS AppDev Studio makes use of the features of TCP/IP to manage Web communication.

---

[2] See Katie Hafner and Matthew Lyon, *Where the Wizards Stay Up Late: The Origins of the Internet* (Touchstone Press, 1998).

The existence of the Internet as a shared resource led to an interest in simplifying the user interface. Clearly, a standardized method for access and display was necessary. In 1989, Tim Berners-Lee, a British computer scientist working at CERN, proposed a global project to allow sharing information over this new medium. He developed the first Web server, using the Hypertext Transfer Protocol (HTTP), and the first Web client. He called the combination of these technologies the *World Wide Web (WWW)*. The World Wide Web first became available on the Internet in the summer of 1991. It is the conjunction of the physical network and the World Wide Web user interface that has led to the enormous growth in Internet connectivity and Web development.

Berners-Lee's brilliant contribution was defining how documents could include embedded *links*, or *hypertext*, which would allow users to connect seamlessly to documents on widely distributed computers. The HTTP standard uses the client/server model previously described. A program running on the server continuously listens for client messages; a second program, called a *Web browser*, runs on the client.

The browser has two jobs. First, it can send messages to the server, correctly encoded in HTTP, using TCP/IP to format and address each message. When the user types the address of a Web server in the browser window, the client sends something like the following request over the Internet to the server at that Internet address:

```
GET /index.html HTTP/1.1
```

The HTTP protocol defines how messages are formatted and transmitted, and what actions Web servers and browsers should take when receiving a request. When the Web server receives a transmission encoded using the HTTP standard, it attempts to respond appropriately. In this example, it responds by sending back the document *index.html* from a specified Web page directory on the server.

In order to find the right server, the local client has to figure out the correct IP address. It does this by sending a preliminary message to a *Domain Name System* (DNS) server with the name of the Web server it has been asked to locate. The DNS server receives this *Uniform Resource Locator* (URL) and replies with a numeric IP address where the Web server can be reached. The browser then inserts this IP address into the header of the outgoing message and transmits it to the requested Web server.

You can also type in an IP address directly as the URL. IP addresses are familiar to most Web users as a set of four three-digit numbers. For example, 66.218.71.81 is the IP address that corresponds to the www.yahoo.com home page. Each of the four fields separated by dots is a number in the range 0 through 255; these are the decimal numbers that can be represented in computer binary language in 8 bits—that is, 2^8 or 256 possible combinations. On most modern computers, 32 bits equals one *word* in storage. Thus four 8-bit numbers were used as the original format of an IP address. As a consequence of the enormous expansion of the Internet, the system is rapidly running out of addresses, and new standards such as IPV6 are currently being advanced to increase the size of possible IP addresses.

The second function the browser provides is the capability to display the received file, using the rules for decoding *Hypertext Markup Language* (HTML) documents. HTML is the set of rules describing the contents of Web files. The development of the HTML standard was what transformed the World Wide Web from an academic curiosity to the ubiquitous entity it is now.

Markup Languages

Standard Generalized Markup Language (SGML) is the standard for organizing the elements of a document. SGML was developed and proposed by the ISO in 1986. This system uses *markup tags* enclosed in angle brackets (<>) to identify and delimit the various parts of a document (header, body, paragraph, and so forth). Although SGML itself is too large and cumbersome to have wide appeal, various subsets of the standard, including HTML and *Extensible Markup Language* (XML) have become tremendously important for international e-commerce. Note that markup languages such as HTML are not programming languages. In form, encoded texts are more like the familiar word processing documents, containing instruction as to how the information contained is to be formatted and displayed. The markup language is simply a set of rules for encoding the text.

XML uses customized tags to provide for verifiable transmittal of data between applications and between organizations. In contrast to HTML, XML was designed to support only the information content of the message; in HTML, this content is combined with the presentation and formatting of the data as well. Most recently, *Extensible Hypertext Markup Language* (XHTML) has been proposed as a way to combine the validation features of XML with HTML formatting capabilities.

Finally, *Dynamic HTML* (DHTML) refers to Web content that can change each time it is viewed. It is important to note that the term is frequently used to refer to two quite different things. The first meaning, which is the one that is used in this book, is simply Web content that can change each time it is viewed.

The second use refers to competing proposals from Microsoft and Netscape to the World Wide Web Consortium (W3C) for various extensions to HTML that allow a Web page to react to user input without sending requests to the Web server. The current position of the W3C on DHTML is as follows:

> "Dynamic HTML" is a term used by some vendors to describe the combination of HTML, style sheets and scripts that allows documents to be animated. The W3C has received several submissions from members companies on the way in which the object model of HTML documents should be exposed to scripts. These submissions do not propose any new HTML tags or style sheet technology. The W3C DOM WG is working hard to make sure interoperable and scripting-language neutral solutions are agreed upon.
> (See "Why the Document Object Model?" http://www.w3.org/DOM/)

Interested users can find more information on DHTML and DOM in the references at the end of this chapter.

There are two main strategies for managing dynamic Web page content:

- *Client-side* DHTML uses JavaScript, cascading style sheets (CSS) or Java applets.
- *Server-side* content can be distributed using Common Gateway Interface (CGI) scripts, PHP: Hypertext Preprocessor, Java servlets, or JavaServer Pages (JSP).

In addition, Microsoft has developed a parallel set of technologies, including JavaScript and *Visual Basic Scripting Edition* (VBScript) which can be used to create *Active Server Pages* (ASP) on the server.[3] These latter all require some version of the Windows operating system, and are explicitly integrated with Microsoft *Internet Information Services* (IIS) and SQL Server, the Microsoft relational database management system.

Since the introduction of SAS 5 in the 1980s, SAS software is and has been largely platform-independent. The SAS AppDev Studio toolkit is available only for Windows, however. Nonetheless, Web pages developed with SAS AppDev Studio can be deployed equally well in the UNIX and mainframe server environments as on the Windows platform, although there are necessarily some differences in implementation. The examples in this book show how to use the tools available for the Linux environment as well as for the Windows XP and Windows 2000 platforms.

Deploying Content on the Web Server

So far, the term "Web server" has been loosely used to mean two quite different things. Strictly speaking, a Web server is not a computer; it is a computer program. Still, most people use the term "server" to refer both to the software and to the hardware on which it runs.

The usual model for a *three-tier* Web computing environment looks something like the following figure:

Figure 1.1 Typical Web Client/Server Configuration

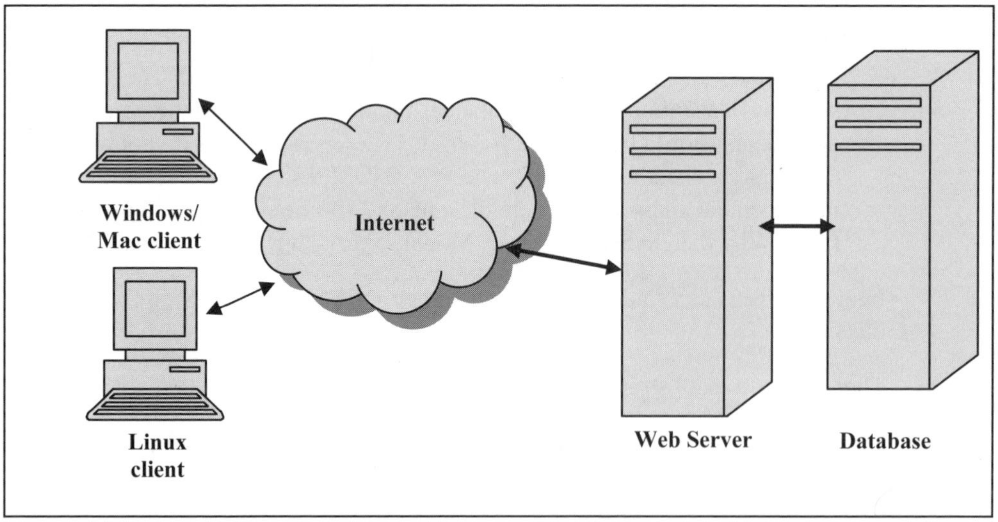

In this design, the client computers are connected to the Internet (via TCP/IP), which in turn is connected to the computer on which the Web server software is running. In addition, the Web server can talk to a database server running on a third computer system.

---

[3] ASP is an older technology from Microsoft; the current platform for delivering active server content is called ASP.NET 2.0; see http://www.asp.net/ for more information.

While this is a common model, in principle all three programs—the client, the Web server, and the database server—could be on two computers, or even all on one. In the previous figure, the data server is located on a different hardware platform from the Web server. In many installations, the two servers are both on the same system. In either case, the SAS/SHARE, SAS/CONNECT or SAS Integration Technologies products must be licensed in order to use SAS as the back-end database service; see Chapter 4, "Remote Access to SAS Data" for more detail on how to implement this connection.

The neat trick about TCP/IP is that it does not know or care where the destination IP address is actually located. The client computer is perfectly happy talking to a Web server that happens to physically reside on the same machine. When you install SAS AppDev Studio on a Windows client, the installation routine offers to install a Web server for you so that you can test your Web pages. The server that currently comes bundled with SAS AppDev Studio is from the Apache Software Foundation (http://www.apache.org).

The Apache HTTP server has been the most popular Web server on the Internet since 1996, with about 70% of the installed server base of 75 million sites (as of December 2005, according to http://www.netcraft.com/survey). Several other Web server programs are available, in particular Microsoft IIS with about a 20% market share,[4] along with various others, none of which has more than 3% penetration.

As a historical note, the Apache HTTP server software was originally developed as a voluntary, part-time project:

> In February of 1995, the most popular server software on the Web was the public domain HTTP daemon developed by Rob McCool at the National Center for Supercomputing Applications, University of Illinois, Urbana-Champaign. However, development of that httpd had stalled after Rob left NCSA in mid-1994, and many webmasters had developed their own extensions and bug fixes that were in need of a common distribution. A small group of these webmasters, contacted via private e-mail, gathered together for the purpose of coordinating their changes (in the form of "patches").
> ("How Apache Came to Be," http://httpd.apache.org/ABOUT_APACHE.html)

Consequently, the project became known as "a patchy" server.

The two main reasons why Apache is so dominant are (1) it works, and (2) it's free. In addition, because there are over 50 million Apache installations worldwide, there is a great deal of free support available from user groups and other resources. The big drawback with Apache (and corresponding advantage for IIS) is that since the former is open source, if it doesn't work you have to figure it out yourself. There is no service number to call when things go wrong (but no service fees, either!).

If you are working in an environment where you have access to a remote Web server, you need to find out from your system administrator whether it is (1) an Apache server on Windows, (2) Microsoft IIS, or (3) an Apache server on UNIX (or Linux). The examples that follow focus on Apache, but IIS is conceptually similar. All Web server programs serve HTML pages from a specific directory; the main difference is just the name of the folder where the pages are located.

---

[4] A recent Microsoft press release indicates that in June 2005 ASP and ASP.NET represented about 44% of the content on Fortune 1000 corporate application servers. This is probably a result of the fact that larger companies are less likely to reply on free software. (http://www.port80software.com/about/press/060105)

Using the Apache Web Server on Windows

Apache was originally written to run on various flavors of the UNIX operating system. While earlier versions of Apache, notably 1.3, came with warnings that the Windows performance was inferior to the UNIX versions, the newest releases (starting with version 2.0) are said to work reliably with current versions of the Windows operating system. You can download a copy of the binary executables for most operating systems from the Apache Web site at http://httpd.apache.org/, or if you are installing SAS AppDev Studio on a client system, the setup routine will ask you if you want to install a copy locally.

Since SAS AppDev Studio is a Windows-only product, the version of the Apache Web server supplied is intended for Windows NT, 2000, and XP. SAS has automated the installation and configuration of the server program so that it is fairly simple to operate in the Windows environment. On a Windows system, you can define Apache as a service, and start and stop it from the **Start ▶ Programs** menu. This is the recommended approach, since that way the Web server will start automatically when you reboot your computer. Just go to **Programs ▶ Apache Web server ▶ Apache as a Service** and select **Install Service**. You can start and stop the service from the same menu.

In order to display HTML content, the pages must first be copied to a specific directory on the Web server system. Under Windows, it is possible to copy HTML documents to the server by mapping a network drive (Z: for example) using Windows Network Neighborhood, and then just dragging or dropping the HTML files to this drive. (As we shall see, this task is somewhat more complex on a UNIX or Linux system.) Even if the Web server is on the same PC you are using, it is still a good idea to map a drive to the directory where you want to display your Web pages.

Unless otherwise requested, the SAS AppDev Studio setup routine will install the Apache 2.0 server in `C:\Program Files\Apache Group\Apache2`. Within the Apache folder, the executable file `apache.exe` is located in the subdirectory `bin`. The default root directory for HTML documents is `htdocs`. Default file locations for Apache are specified in the configuration directory `conf\httpd.conf`. Unless you know what you are doing, you should not try to change these yourself. On a corporate Web server, you almost certainly will not have permission to edit this file; if you have installed Apache on your local workstation, you probably do not need to change the defaults anyway.

As noted above, the folder `C:/Program Files/Apache Group/Apache2/htdocs` is the default Apache document root directory. Unless you tell it otherwise, Apache will try to serve Web pages from this folder. You do not need to (nor should you) try to specify "htdocs" in the URL. If the name of the Web page is `example.htm`, located in the `htdocs` folder, the URL for this Web page would be http://<server-name>/example.htm.

At most sites, users do not have write access to the `htdocs` folder on the Web server. In this case, the user has to contact the system administrator for a directory structure with the correct access and permissions. The URL would thus contain an alias to this folder—for example http://<server-name>/~username/example.htm. If you get the nasty message "The page you are looking for is currently unavailable," the HTML file is most likely not in the right directory. This would be a good time to get help from someone who has tried this before on your Web server.

If your Web server is on a different network, including one on the other side of the world, it is still possible to map a drive locally using *WebDAV* (Web-based Distributed Authoring and Versioning). If your server is configured to support Web DAV, you can set up a client PC to access a Web directly using HTTP. Although it is technically possible to do this using Web Folders in Microsoft Office or Internet Explorer, most people prefer to use a dedicated client application such as NetDrive or WebDrive; see the resources listed at the end of this chapter for information about these products. Using WebDAV greatly simplifies the process of managing Web content. Check with your system administrator to find out whether WebDAV is supported in your environment.

Using the Apache Web Server on UNIX/Linux

In a Windows environment, as long as you have permission to write to the public documents folder, you can just map a drive to this folder and drag and drop your Web pages there. On UNIX systems, transferring documents is slightly more complex. The default directory paths are specified in the file **conf/httpd.conf** under the server root directory. As noted above, these values must be supplied at installation time by the system administrator. If you are the system administrator and you installed Apache yourself, change to the Apache directory and try typing the following command:

```
grep DocumentRoot conf/httpd.conf
```

This should list out the line in the file containing the value assigned to this parameter. One common location for sites running Apache 2 is **/usr/local/Apache2/htdocs**. Earlier versions of Apache such as 1.3 used **/var/www/html**, and the configuration files were in **/etc/apache2/conf**.

The UNIX system administrator has to set up a password and some space on the Web server for each user. Ask what directory you should use for your Web pages, and whether you should use File Transfer Protocol (FTP) or Secure File Transfer Protocol (SFTP) to transfer them. FTP is the TCP/IP protocol for copying files from one computer to another. You can use it for copying files from one UNIX system to another, or from a Windows system to a UNIX server. FTP is a relatively old protocol. Unfortunately it has one major security problem. When you type in your user name and password, they are sent over the network unencrypted. This is bad enough when connecting to an FTP server on your LAN or Intranet, but it is a real no-no when sending a file over the Internet to a remote server. Anyone can find out your password just by monitoring the network traffic. Consequently most sites are now requiring SFTP. This is easy to set up, but again, you need to talk to your system administrator about what you need to do at your specific installation.

In either case, the syntax is easy, if you just follow these steps:

1. On a Windows client, open an MS-DOS window. On a UNIX system, open a terminal window. In either case, you should have a prompt character after which you can type commands.
2. Open a connection to the remote system by typing **ftp *host-name*** or **ftp *host-name***, where ***host-name*** is the name or IP address of the remote computer.
3. You will be prompted for your user name and password. Use the ones you got from your system administrator.
4. You may need to change to your directory. Type **cd *name***, where ***name*** is the path to the directory where you want to put your HTML pages.
5. To transfer the files, type **put *name***, where ***name*** is the name of the HTML document you want to display.
6. Type **quit**.

Note that a variety of GUI-based clients support FTP and SFTP; using these allows you to drag and drop files to server directories, assuming your user ID has the proper permissions to do so.

You should now be able to open a Web browser on your PC and type in the URL of the document you have just copied. Just as with the Windows version, this URL will consist of the name of the server (which you found out from the system administrator), followed by any specific directory locations (like **sasweb**), followed by the name of the HTML page you want to display.

Using Microsoft Internet Information Server

The Microsoft Windows Server and Windows XP Professional editions come with IIS included. This product does not run under UNIX, but for sites with Windows servers, it is an ideal choice. The Web server can be installed in a few minutes from the operating system installation CD, or by going to **Start ▶ Control Panel ▶ Add or Remove Programs ▶ Add/Remove Windows Components**.

The default installation directory for IIS is **C:\InetPub**; the folder **wwwroot** is the root directory for serving Web pages. Once IIS has been installed on the server, you can get detailed instructions on use by opening http://<server-name>/iishelp, where <server-name> is the host name for your Web server.

Whether your server is running Apache or IIS, locally on your PC or in Australia, the idea is the same. There will be one specific directory on the server where you want to copy your HTML documents. The URL to this directory will consist of the server name, possibly followed by the path to your folder, followed by the name of your HTML document.

Now that you know how to deploy Web pages, it is time to start creating a few. The following chapter is a short introduction to using HTML to create Web pages. If you are familiar with HTML, you may want to go directly to Chapter 3, "Creating Static HTML Output," which covers several options for creating static Web pages with SAS.

References

SAS Publications
URL references are current as of the date of publication.

- SAS Institute Inc. 2001. *Getting Started with AppDev Studio.* 2nd ed. Cary, NC: SAS Institute Inc.
- SAS Institute Inc. 2001. *SAS Web Tools: Overview of SAS Web Technology* (Course Notes). http://support.sas.com/training/us/crs/wovr.html

Web Programming
As of December 2005, there were 2200 volumes on the topic "Web programming" at http://www.amazon.com. The following are the works cited in the text.

- Cooper, Alan. 2004. *The Inmates Are Running the Asylum: Why High Tech Products Drive Us Crazy and How To Restore The Sanity*, 2nd ed. Indianapolis, IN: Sams.
- Hafner, Katie and Matthew Lyon. 1998. *Where the Wizards Stay Up Late: The Origins of the Internet.* New York, NY: Touchstone Press.
- Krug, Steve. 2005. *Don't Make Me Think: A Common Sense Approach to Web Usability.* 2nd ed. Indianapolis, IN: New Riders Press.

- Nielson, Jakob. 2000. *Designing Web Usability*: *The Practice of Simplicity*. Indianapolis, IN: New Riders Press.
- Torvalds, Linus and David Diamond. 2001. *Just for Fun: The Story of an Accidental Revolutionary*. New York, NY: HarperBusiness.

Links

- Apache HTTP Server Project – http://httpd.apache.org/ABOUT_APACHE.html
- Document Object Model – http://www.w3.org/DOM/
- Internet Protocols – http://directory.google.com/Top/Computers/Internet/Protocols
- Microsoft Internet Information Services (IIS) – http://www.microsoft.com/WindowsServer2003/iis/
- Novell NetDrive 4.1 – http://www.novell.com/documentation/ifolder21/netdrive/data/a2iii88.html
- South River Technologies WebDrive – http://www.webdrive.com/products/webdrive/
- WebDAV – http://www.webdav.org/
- Web Server Survey – http://www.netcraft.com/survey
- Working with Distributed Authoring and Versioning (DAV) and Web Folders – http://support.microsoft.com/kb/q221600/

Chapter 2

Introduction to HTML

Hypertext Markup Language

There are many good how-to books on learning HTML; a selection of titles is included in the references at the end of this chapter. Since this book is mostly about SAS AppDev Studio, the discussion of HTML features is confined to those that are likely to be most useful to the SAS user. Specifically these include tables, forms and images, all of which are used by the Output Delivery System (ODS) to format and display output (see Chapter 3, "Creating Static HTML Output," for an introduction to ODS). If you are already familiar with using these tools, feel free to skip this chapter.

All markup languages use *tags* to annotate the document content. In HTML, there is a relatively short list of standard tags that you need to learn. Once these are mastered, it is simply a question of assembling them into the desired order. Of course in practice it is rarely so simple, but the general principles are not difficult to learn.

Writing HTML can be very tedious, since there is necessarily a lot of repetition. For this reason, a number of tools are widely available to automate the process. These are generally referred to as IDEs (for *Integrated Development Environment* or *Integrated Design Environment*). The webAF software that ships with SAS AppDev Studio is such an IDE, as we shall see later on in this book.

However, many professional Web developers avoid relying on formal IDEs, preferring instead to construct HTML from scratch using a text editor such as vi, Notepad, KEDIT,[1] or TextPad[2].

A Web page is just an ASCII text file, with some markup tags inserted to format and display the text. Some of the most common markup tags are listed in Table 2.1:

Table 2.1 Commonly Used HTML Tags

Tag	Function
<a href="URL">text</a>	Insert an anchor tag, link to the URL
<blockquote>text</blockquote>	Indent the text one tab position
 	Insert a new line (no closing tag)
<div>text</div>	Separate out a block of text for formatting
italic	Emphasize the text
<hr />	Insert a single horizontal line (no closing tag)
<img />	Insert a graphic image or picture (no closing tag)
<li>list element</li>	Bullet or number the item, keep the text together (The item must be enclosed in a numbered or bulleted list.)
<ol>ordered list</ol>	Number the items in the list
<p>paragraph</p>	Keep the text together, new line at the end
<pre>preformatted text</pre>	Maintain the original formatting, including white spaces and line breaks
text	Separate out a segment of text within a larger block for formatting
bold	Create bold text
<table>table</table>	Insert a table
<td>table cell</td>	Define the beginning and end of a table cell
<th>column heading</th>	Define the beginning and end of a table header cell
<tr>table row</tr>	Define the beginning and end of a table row
<u>underline</u>	Create underlined text
<ul>unordered list</ul>	Bullet the items in the list

The best way to understand how to use these tags is by example; Display 2.1 shows a sample HTML page as it would appear in a Web browser.

[1] KEDIT is a trademark of Mansfield Software Group, Inc.
[2] TextPad is a trademark of Helios Software Solutions.

Display 2.1 Sample Web Page

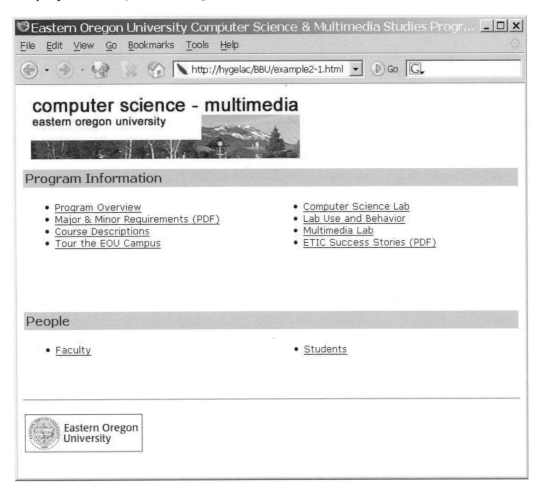

Clicking on **View ▶ Source** in the browser menu displays the following page source code for the page `index.html`. This code may appear intimidating at first, but as noted previously it is constructed simply by repeating standard elements; see the discussion in the following section for an explanation of the markup.

Example 2.1 Sample HTML Page Source

```
<!DOCTYPE html PUBLIC "-//W3C//DTD XHTML 1.0 Strict//EN"
    "http://www.w3.org/TR/xhtml1/DTD/xhtml1-strict.dtd">
<html public="public" xhtml="xhtml"
  xmlns="http://www.w3.org/1999/xhtml">
<head>
  <title>Eastern Oregon University Computer Science &amp;
     Multimedia Studies Program</title>
  <link href="styles.css" type="text/css" rel="stylesheet" />
</head>
<body>
<div><img class="hdr" src="http://cs.eou.edu/images/csmm.jpg"
  alt="EOU Computer Science/Multimedia Studies Program" />
</div>
<table summary="Program Information">
```

```
<tr>
  <td colspan="2" class="row_hdr">Program Information</td>
</tr>
<tr>
  <td class="row1">
  <ul>
      <li><a href="programinfo.htm">Program Overview</a></li>
      <li><a href="http://www.eou.edu/advising/checklist/CSMM.pdf">
           Major &amp; Minor Requirements (PDF)</a></li>
      <li><a href="http://www.eou.edu/catalog/compmulti.html">
           Course Descriptions</a></li>
      <li><a href="http://cs.eou.edu/CSMM/EOU_tour_site/index.html">
           Tour the EOU Campus</a></li>
  </ul>
  </td>
  <td class="row1">
  <ul>
      <li><a href="cslab.htm">Computer Science Lab</a></li>
      <li><a href="http://cs.eou.edu/cslabuse.htm">
               Lab Use and Behavior</a></li>
      <li><a href="mmlab.htm">Multimedia Lab</a></li>
      <li><a href="http://www.oregonetic.org/
           05-07/ETICSuccessStoriesRevC.pdf">
           ETIC Success Stories (PDF)</a></li>
  </ul>
  </td>
</tr>
<tr>
  <td colspan="2" class="row_hdr">People</td>
</tr>
<tr>
  <td class="row2">
  <ul>
      <li><a href="http://cs.eou.edu/faculty.htm">
           Faculty</a></li>
  </ul>
  </td>
  <td class="row2">
  <ul>
      <li><a href="http://cs.eou.edu/students.htm">
           Students</a></li>
  </ul>
  </td>
</tr>
</table>
<hr />
<p><a href="http://www.eou.edu">
  <img class="logo"
       src="http://cs.eou.edu/images/eoulogow2.gif"
       alt="EOU logo" /></a></p>
</body>
</html>
```

HTML vs. XHTML

The World Wide Web Consortium (W3C), the standards body responsible for HTML, advocates using "a stricter and cleaner reformulation" called *Extensible Hypertext Markup Language* (XHTML):

> XHTML is a family of current and future document types and modules that reproduce, subset, and extend HTML 4. XHTML family document types are XML based, and ultimately are designed to work in conjunction with XML-based user agents.
> (http://www.w3.org/TR/xhtml1/)

The main difference between XHTML and HTML is the requirement that XHTML documents be "well-formed." This means that all elements must be in lower case, have closing tags, and nest properly. In addition, attributes must be quoted. It turns out that it's not very difficult to make your HTML pages meet these requirements; in fact, it's good practice. The examples shown in this chapter use the XHTML conventions, but that should not distract from the essential information.

The most important difference between HTML and XHTML is that the latter must conform to a *Document Type Definition* (DTD) for XML-based Web pages. What this means in practice is that you cannot use `<font>` tags or other formatting instructions in your pages. In the 15 years that the Web has been around, developers have recognized the importance of separating the *appearance* of the page from the *content*. The XHTML standard requires that all page formats must be included in style sheets, not as part of the markup. The big advantage of this requirement is that it enhances internationalization, as well as making it much easier to achieve a consistent look and feel across your site.

A Web browser uses the markup tags to determine how to display the document. All tags are constructed in the same way. Each tag is surrounded by angle brackets. For example, the page shown in Example 2.1 begins with a `<!DOCTYPE>` tag. This tells the browser which HTML or XHTML specification is used in the document, in this case strict XHTML:

```
<!DOCTYPE html PUBLIC "-//W3C//DTD XHTML 1.0 Strict//EN"
    "http://www.w3.org/TR/xhtml1/DTD/xhtml1-strict.dtd">
```

In the XHTML standard, the enclosing tag for the Web page, known as the root element, must be `<html>`, as shown in the second line:

```
<html public="public"
   xhtml="xhtml"
   xmlns="http://www.w3.org/1999/xhtml">
```

The tag shown includes additional information that is required for XHTML, in particular the *namespace* that is used. In this case, the page is using the public namespace `xmlns` (*XML name space*). Namespaces are a way of defining groups of tags; for more information, see the resources listed at the end of this chapter.

All XHTML tags must have a closing tag to set off the included text. These are just like the opening tags, except that they begin with `</` rather than just `<`. The very last tag at the end of the document is the HTML closing tag `</html>`, which balances the opening tag and closes the document.

The next section of the preceding example shows how tags can be nested. The `<head></head>` tags identify the section of the document containing the page header. A `<title>` tag specifies the text that should appear in the title bar (not on the Web page itself). The `</title>` tag closes the text string, and then the `</head>` tag closes the header. Other information that can go in the HTML head includes `<style>` tags, used to create cascading style sheet (CSS) styles (see the following section) and `<meta>` tags, which convey information to Web search engines.

Note that HTML 4.0 tags are not case sensitive: you can say `<HEAD>`, `<head>`, `<Head>`, or even `<hEaD>`. The standard, however, is that all tags should be in lower case. The HTML standard does not specify the order of the closing tags, so long as all open tags get closed. As noted previously, allowing unclosed tags is perceived as a defect in the HTML specification that has been remedied in the newer XHTML specification. In addition, like SAS, HTML ignores white space. You can indent your page source code any way that seems helpful, and you can use comments with a spendthrift hand.

The `<link>` tag is used to include another Web source document[3]. This tag can appear only in the <head> section of a document and cannot have any included text. In the example, it is included to specify a link to a user-defined CSS called **styles.css** in the current directory. (See the discussion of style sheets in the next section for an explanation of how this works.)

The HTML specification indicates that tags may also have associated attributes. Attributes consist of pairs of standard names and values, separated by an equals sign. Use quotes around attribute values; the current HTML standard does not require them, but XML and XHTML do, and it is a good idea to get into the practice. In this case, three attributes are used on the `<link>` tag:

- `href` – location of the source document to be included (required)
- `type` – content type, specifies the style sheet language
- `rel` – relationship, specifies link type

The main section of this document is delimited by the tags `<body>` and `</body>`. Logically enough, these enclose the body of the document, the content displayed in the browser window.

The example illustrates the XHTML convention that all tags must be enclosed in an outer block. In the example, a `<div>` tag is used to set off an `<img>` tag. Image tags, as the name suggests, are used to include pictures in HTML documents, as the example illustrates:

```
<img class="hdr"
   src="http://cs.eou.edu/images/csmm.jpg"
   alt="EOU Computer Science/Multimedia Studies Program" />
```

The `<img>` tag differs from those described above in that there is no corresponding closing tag; all of the necessary information is conveyed by the attributes included in the tag. It is consequently considered good form to indicate that this is an *empty* tag by using `/>` as the closing symbol.

---

[3] See http://www.w3.org/TR/html4/struct/links.html for more information about links.

Three attributes are used on the `<img>` tag:

- `class` – a format defined in the accompanying style sheet
- `src` – a link to the image displayed (required)
- `alt` – alternate text that appears if the image cannot be displayed (required)

SAS uses `<img>` tags to display the output of SAS/GRAPH procedures. Images can be generated in either of two Web-compatible forms:

- GIF *(Graphics Interchange Format)* – an older image format usually used for drawings and icons
- JPEG *(Joint Photographic Experts Group)* – a compressed format that is used to display photographs and other images with a lot of detail. On DOS systems, the extension .jpg was used since the operating system did not support the four-letter extension .jpeg. Windows and UNIX systems can recognize either.

In SAS 6 the GOPTIONS statement could be used to redirect graphical output to a GIF file. In SAS 7 and subsequent releases, the SAS Output Delivery System can now create image files from SAS/GRAPH output (see Chapter 3, "Creating Static HTML Output").

Tables can be an extremely useful formatting technique, of which ODS makes good use. Example 2.1 includes a table that has four rows and two columns, delimited by the opening and closing `<table></table>` tags. The `<table>` tag illustrated includes a user-defined label for the table as an optional attribute:

```
<table summary="Program Information">
```

The `<tr>` and `</tr>` tags delimit table rows whereas `<td>` and `</td>` enclose the cell contents; these tags can appear only nested inside of `<table></table>` elements. Note in the example that each row of this table has two columns, so there are two sets of `<td>` tags for each row.

The first row of the table has only a single cell, spanning both columns:

```
<tr>
    <td colspan="2" class="row_hdr">Program Information</td>
</tr>
```

This tag uses a `class` attribute to identify the presentation format from the included style sheet; see the following discussion on CSS.

The contents of a table cell are not limited only to data values. Tables can hold images, multimedia clips, and even other tables. Web designers often employ nested tables in order to achieve more control over page layout, although current practice is to use `<div>` tags and styles to achieve the same end.

The second row of the table contains two bulleted lists, one in each column. HTML has three kinds of lists. See http://www.w3.org/TR/html401/struct/lists.html for more details about HTML lists.

- unordered – bulleted
- ordered – numbered
- definition – a series of term/definition pairs

The elements in each type of list are enclosed in `<li>` tags.

The lists include links to other files, as indicated by *anchor* tags <a>. The *href* attribute on an <a> tag indicates that the attribute value is to be treated as a hypertext reference or link.

Two kinds of links are shown: relative and absolute. An anchor tag like the following indicates that the referenced document is on the same Web server as the one displayed, in this case in the same directory.

```
<a href="programinfo.htm">Program Overview</a>
```

Clicking on the tag causes that file to be loaded into the memory of the client computer and displayed in the browser window.

Absolute tags work the same way, except that they are qualified with the name of the Web server. For example, the following tag displays the message "Major & Minor Requirements (PDF)" as a link to the indicated file on the EOU main server.

```
<a href="http://www.eou.edu/advising/checklist/CSMM.pdf">
  Major &amp; Minor Requirements (PDF)
</a>
```

The string & is an *entity code*. Since the ampersand is a special character in HTML, it must be written as an entity, as shown. See http://www.ascii.cl/htmlcodes.htm for a complete list of the codes corresponding to various ASCII characters.

The <hr> tag inserts a dividing line on the page. Since this tag has no end tag, the convention used is to include a slash before the concluding bracket, as shown.

The last section of the example message body consists of a *paragraph*, delimited by <p> and </p> tags. This paragraph illustrates the use of an <img> tag for a hyperlink. Clicking on the picture takes you to the URL provided. Note that since the opening <a> tag occurs after the opening <p> tag, the closing </a> tag must occur before the closing </p> tag. This is what the W3C refers to as proper nesting.

Finally, the </body> tag closes the description of the body of the document, and the </html> tag closes the document itself.

Formatting HTML Using Style Sheets

Style sheets are used to create a uniform appearance for a Web site by separating the page content from the presentation. They also simplify Web page maintenance. When the marketing folks decide at the last minute that the shade of green on all of the Web page headers isn't quite right— they'd like a darker, more forest green—the clever designer, who coded the heading color on a style sheet, can make the change in one place rather than separately on every Web page.

In office automation suites such as Microsoft Office, style sheets, also known as *templates,* are maintained as separate files or forms that specify the document layout. Web developers can use the same principal, storing page formats and other style information in a separate file on the server. The World Wide Web Consortium (W3C) has developed a standard for HTML called *cascading style sheets* or CSS. A single file with the extension .css can be used to apply a common format to many Web pages. The term "cascading" derives from the fact that multiple style sheets can be applied to the same Web page, so that individual users can customize organizational formatting rules by appending their own styles after the standard ones.

It also is possible to code user-specified formats directly into an HTML document, using the `<style></style>` tags. Styles specified in these tags share a common syntax with external CSS files. The advantage of *external* style sheets is that one style can be used to format many different HTML pages. With inline formatting, a global change requires editing all of the affected pages.

If you are using SAS to create output pages, you can specify a CSS file to format your pages according to the rules at your site or your own personal preferences. A number of tutorials are available on CSS. Type "cascading style sheets" into your favorite search engine and you will be directed to a good many of them. See http://www.w3.org/Style/CSS/ for a list of useful references.

The following simple example illustrates how to create and link an external style sheet to an HTML page. First, create a text file (in your favorite editor) that contains just the formatting instructions. The syntax of CSS requires an element name followed by a set of style instructions, separated by semicolons, all enclosed in curly braces { }:

Example 2.2 Sample Cascading Style Sheet

```
body{ background-image: url("underwater.jpg") }
h1{ color: red; font-weight: bold; text-align: center }
h2{ color: blue; font-weight: bold; text-align: center }
p{ color: yellow; font-size: 14 pt; text-align: left }
```

These instructions specify that the image file **underwater.jpg** should be used as the background. Headings are in bold and centered. Main titles (`h1`) are in red; sub-heads (`h2`) are blue. Paragraph text is black and the font size is 14 point.

Save this file with the extension .css (for example, as **mystyle.css**) and add a link to the new file in the header of your HTML document:

Example 2.3 Sample Page with Link to External CSS

```
<html>
<head>
  <link rel="stylesheet" type="text/css" href="mystyle.css">
</head>
<body>
  <h1>heading 1</h1>
  <h2>heading 2</h2>
  <p>paragraph1</p>
  <p>paragraph2</p>
  <p>paragraph3</p>
</body>
</html>
```

This will cause the specified format to be applied to any document that contains this `<link>` tag. The following example shows the resulting HTML page.

Display 2.2 Sample Output Formatted with CSS

The example illustrated in Display 2.1 also includes a link to an external style sheet. As noted above, using a single style sheet to format multiple Web pages allows the developer to create a consistent look across the site. The included style sheet is called **styles.css** as shown in Example 2.4 below:

Example 2.4 Sample Cascading Style Sheet with Table Formats

```
body {
        background-color: #FFFFFF;
    font-family: verdana, helvetica, arial;
}
:link {
    color: #000080;
}
:visited {
    color: #0000FF
}
:active {
    color: #800000
}
table {
    width:100%;
```

```
        height:300px;
        border: 0;
    }
    td {
        width: 50%;
        font-size: 10pt;
        vertical-align: top;
        padding: 2pt;
    }
    td.row_hdr {
        color: #800000;
        background-color: #C0C0C0;
        font-size: 12pt;
        font-weight: bold;
        height: 20px;
        width: 100%;
    }
    img.hdr         {
        height: 100px;
        width:  400px;
    }
    img.logo {
        border: 0;
        float:      left;
        height: 55px;
        width: 165px;
    }
```

This document is quite similar to Example 2.2, with the addition of a few more formats.
Altogether, nine different values are defined in this style sheet:

- `body` – applies to all content between the `<body>` and `</body>` tags
- `:link` – defines a link style (or *pseudoclass*) for hyperlinks that have not been visited—
  that is, pages that are not in the browser cache
- `:visited` – describes any hyperlink that has been cached
- `:active` – describes a hyperlink that has been clicked but not yet released
- `table` – applies to all content between the `<table>` and `</table>` tags
- `td` – applies to all table cells
- `td.row_hdr` – defines a style that applies only to table cells labeled with
  `class=row_hdr`
- `img.hdr` – applies only to images labeled with `class=hdr`
- `img.logo` – applies only to images labeled with `class=logo`

For more information about formatting pages with style sheets, see the references at the end of
this chapter.

Including User-supplied Information with Forms

HTML *forms* provide a way to collect user input. Most people are familiar with Web pages that
are used to log on to services or order products; these are usually managed by using forms.
Display 2.3 shows a sample form with a variety of controls:

Display 2.3 Sample HTML Form

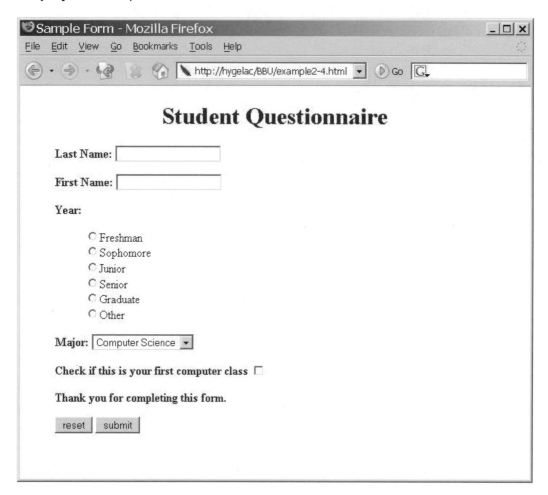

The source code for this form is shown in Example 2.5:

Example 2.5 Sample HTML Source Code

```
<!DOCTYPE html PUBLIC "-//W3C//DTD XHTML 1.0 Strict//EN"
    "http://www.w3.org/TR/xhtml1/DTD/xhtml1-strict.dtd">
<html xmlns="http://www.w3.org/1999/xhtml">
<head>
  <title>Sample Form</title>
  <style type="text/css">
      h1 { text-align: center; color: #0000FF }
      .label { font-weight: bold; color: #0000FF }
  </style>
</head>
<body>
<h1>Student Questionnaire</h1>
<form name="students" method="get" action="index.html">
<blockquote>
  <p>
      <span class="label">Last Name:</span>
      <input type="text" name="lastname" />
  </p>
  <p>
```

```
        <span class="label">First Name:</span>
        <input type="text" name="firstname" />
  </p>
  <p>
        <span class="label">Year:</span>
  </p>
  <blockquote>
        <input type="radio" name="year" value="1" />Freshman<br />
        <input type="radio" name="year" value="2" />Sophomore<br />
        <input type="radio" name="year" value="3" />Junior<br />
        <input type="radio" name="year" value="4" />Senior<br />
        <input type="radio" name="year" value="5" />Graduate<br />
        <input type="radio" name="year" value="9" />Other
  </blockquote>
  <p>
        <span class="label">Major:</span>
        <select name="major">
              <option value="CS">Computer Science</option>
              <option value="B">Business</option>
              <option value="F">Forestry</option>
              <option value="M">Media Arts</option>
              <option value="O">Other</option>
              <option value="U">Undecided</option>
        </select>
  </p>
  <p>
        <span class="label">Check if this is your first computer class</span>
        <input type="checkbox" name="firstCS" value="Y" />
  </p>
  <p>
        <span class="label">Thank you for completing this form.</span>
  </p>
  <p>
        <input type="reset" value="reset" />
        <input type="submit" value="submit" />
  </p>
  </blockquote>
  </form>
  </body>
  </html>
```

As in the previous HTML example, this Web page starts out with an `<html>` tag followed by the `<head>` and `</head>` tags enclosing the title bar information. In this case, the style sheet information is local to the page:

```
<style type="text/css">
  h1 { text-align: center; color: #0000FF }
  .label { font-weight: bold; color: #0000FF }
</style>
```

The first instruction indicates that the heading at the start of the body of the document is centered and displayed in blue. The second format is a user-defined one that can be referenced with the name `label`. The initial period indicates that this format is not restricted to a particular element but may be used anywhere.

The reader is here cautioned to take a deep breath, since the description is now going to get a little hairy. The following statement opens a new HTML form:

```
<form name="students" method="get" action="index.html">
```

Forms are used to collect input from the user. Usually a `<form>` tag would have at least two attributes in addition to the optional *name*:

- *action*, which supplies the name of the program to be run when the form is submitted
- *method*, either GET or POST, which tells the browser how to send the form data to the server

When the form is submitted, the browser runs the script or program specified by the *action* attribute on the form. The script can either be local (in VBScript or JavaScript) and executed by the browser, or a server-side script, such as a CGI program or a Java servlet. The script can reference objects on the form based on the *name* attributes on the form tags. How these work, and when to use them, will be described in subsequent chapters. At this point it is sufficient to note that the sample form shown only gathers data; it does not actually do anything with it.

The rest of the page describes the contents of this form, down to the closing `</form>` tag, and concludes with `</body>` and `</html>`. The labels associated with each of the input fields are formatted using the CSS `label` style defined in the document header. A `<span>` tag is used to limit the formatting only to this text.

```
<span class="label">Last Name:</span>
```

The first form control (or *widget* in SAS/AF terminology) is the `<input>` tag. The HTML standard currently defines eight different kinds of inputs, specified by the *type* attribute that appears on the tag:

- Text – a one-line input string
- Password – text that is not displayed when enterned
- Checkbox – a yes or no choice
- Radio – a yes or no choice, but only one can be selected
- Hidden – a field that is not displayed, used to pass information to the server
- Reset – a button that clears all fields on the form
- Submit – a button that sends the form to a Web server
- Image – the same as Submit except it displays a picture instead of a button

The first two input boxes on the form ("lastname" and "firstname") store the text that the user enters in fields specified in the name attribute. An `<input type="text">` tag only allows the user to enter one line of text; pressing the ENTER key goes to the next form control instead of inserting a newline character. In order to input more than one line of text, it is necessary to use the `<textarea>` tag.

The third form control, "year", gets its values from the radio buttons. With this type of control only one choice can be selected. Note that form variables are not typed; you can enter characters or numbers. All values are stored as text, even if the quotes around the values are omitted, as in this example.

The `<select>` tag is an alternative to the `<input>` tag; it allows the user to choose from the list specified by the `<option>` tags. As with the radio buttons and the checkbox, the `value` attribute determines the test string returned from the control.

An input box of type `checkbox` allows the user to select a single value; the *value* attribute is the text string returned when the box is checked.

Finally, the `submit` and `reset` input types display command buttons; the label on the button is determined by the `value` attribute. As noted previously, `reset` clears the form and `submit` sends it to the server. There can be only one submit button and one reset button on a form.

To summarize, HTML tags are used to indicate the layout of the Web page. In addition, `<style>` tags or style sheets can be used to specify generic formats that can be applied to specific page items. The `<table>` tag allows data to be displayed in a tabular format, whereas the `<form>` tag is used to collect information from the user. Pictures and animations can be inserted into the page using `<img>` tags.

References

HTML

URL references are current as of the date of publication.

- Beginners Guide to HTML – http://archive.ncsa.uiuc.edu/General/Internet/WWW/HTMLPrimer.html (A classic)
- Castro, Elizabeth. 2002. *HTML for the World Wide Web with XHTML and CSS: Visual QuickStart Guide*. 5th ed. Berkeley, CA: Peachpit Press.
- Lissa Explains it All: The first and original HTML help JUST for kids – http://www.lissaexplains.com/basics.shtml (This site was developed by a 12-year-old girl.)
- Musciano, Chuck and Bill Kennedy. 2002. *HTML & XHTML: The Definitive Guide*. 5th ed. Sebastopol, CA: O'Reilly & Associates.
- Powell, Thomas. 2003. *HTML & XHTML: The Complete Reference*. 4th ed. Emeryville, CA: McGraw-Hill/Osborne Media.
- W3C CSS Home Page – http://www.w3.org/Style/CSS/
- W3C HTML Home Page – http://www.w3.org/MarkUp/
- W3C XHTML Home Page – http://www.w3.org/TR/xhtml1/

CSS

- Meyer, Eric A. 2004. *Cascading Style Sheets: The Definitive Guide*. 2nd ed. Sebastopol, CA: O'Reilly & Associates.
- Shea, Dave and Molly E. Holzschlag. 2005. *The Zen of CSS Design: Visual Enlightenment for the Web*. Berkeley, CA: Peachpit Press.

Chapter 3

Creating Static HTML Output

Introduction

In the earliest days of the World Wide Web, all content was static. That is, there was no way for a user to modify the appearance or contents of a Web page viewed in the browser window. This worked fine for the original hypertext model, designed for information display only, but it was obviously inadequate for electronic commerce. Later sections of this book introduce the tools for creating dynamic HTML Web pages. In this chapter, however, the focus is on programs that simply generate an HTML file for display by the Web browser.

The examples in this chapter assume that you would like to capture the output from a SAS procedure to display as a Web page. Since this is static HTML, you will need to replace the HTML source each time the values change. If the data is retrospective and unchanging, however, this approach is convenient. You only have to create the output file once, by hand, and then copy it to your Web server directory for display.

An example of static vs. dynamic HTML pages would be two standard reports: one produced periodically and another that relies on results generated in real time. The information in a quarterly or annual report is fixed; the user is not expected (or allowed) to change it. A year-to-date sales report, on the other hand, requires the ability to summarize results dynamically. As noted in the previous chapter, the only way to collect information using HTML forms is to run some kind of program or script when the **Submit** button is pressed. Static HTML is more than adequate for disseminating scientific articles or brochures, but it would not be very effective for making airline reservations.

Note that there are a number of advantages to providing static HTML content. Since they do not require running any additional programs other than the HTTP server, static HTML pages can be served very quickly. At the same time, development effort is minimized. The main drawback is that in order to change the information displayed, the page must be recreated and recopied to the server. As noted previously, it is the responsibility of the designer to determine which data is likely to be static and which dynamic. Effective Web site design requires using the right tool for the right job, and that can mean setting up standardized reports for information that changes infrequently or not at all.

For purposes of comparison, all of the examples in this chapter feature the same program, the following simple TABULATE procedure. The Base SAS installation disks include a number of sample data sets in the default SASHELP library. This sample data is used for the examples in this chapter. Consequently, they can be run at any site without obtaining or installing any special data sets.

When a SAS program is run in batch mode, two output files are created by default: the `.log` file shows the execution time messages and any error messages, while the `.lst` file contains the results. Example 3.1 shows a sample program, and Display 3.1 shows the resulting plain vanilla `.lst` file when the program is run from the command line. Readers unfamiliar with the syntax of PROC TABULATE are referred to the SAS documentation.[1]

The following code is reasonably self-evident; it uses the SASHELP.RETAIL data set to generate various statistics on retail sales by year for a hypothetical company.

Example 3.1 Sample PROC TABULATE Program

```
options noovp nodate nonumber nocenter ls=80;

proc tabulate data=SASHELP.RETAIL
    formchar="|----|+|---+=|-/\<>*";

    title "Retail Sales In Millions Of $";
    class YEAR/descending;
    var SALES;
    table YEAR="" all="Total", SALES="" *
        (sum="Total Sales"*f=dollar8.
```

---

[1] See especially Lauren E. Haworth, *PROC TABULATE By Example* (Cary, NC: SAS Institute Inc., 1999).

```
        pctsum="Overall Percent"*f=8.2
        n="Number of Sales"*f=8.
        mean="Average Sale"*f=dollar8.2
        min="Smallest Sale"*f=dollar8.
        max="Largest Sale"*f=dollar8.)/
   box="Year" rts=8;

run;
```

Most of the program code in this example consists of labels and formats to get the page to display nicely. The FORMCHAR= option, for example, is included to ensure that the resulting output uses the standard ASCII form delimiters, rather than the SAS defaults, which do not work in a Web browser. Without a FORMCHAR= option specified, the output is not useable. The resulting text is formatted for a line printer, as shown below.

Display 3.1 Sample PROC TABULATE Output

```
Retail Sales In Millions Of $
-----------------------------------------------------------------------
|Year  | Total  |Overall | Number |Average |Smallest|Largest |
|      | Sales  |Percent |of Sales|  Sale  |  Sale  |  Sale  |
|------+--------+--------+--------+--------+--------+--------|
|1994  |  $1,874|    5.97|       2| $937.00|    $876|    $998|
|------+--------+--------+--------+--------+--------+--------|
|1993  |  $3,578|   11.40|       4| $894.50|    $758|    $991|
|------+--------+--------+--------+--------+--------+--------|
|1992  |  $3,204|   10.21|       4| $801.00|    $692|    $889|
|------+--------+--------+--------+--------+--------+--------|
|1991  |  $2,947|    9.39|       4| $736.75|    $703|    $807|
|------+--------+--------+--------+--------+--------+--------|
|1990  |  $2,734|    8.71|       4| $683.50|    $606|    $749|
|------+--------+--------+--------+--------+--------+--------|
|1989  |  $2,592|    8.26|       4| $648.00|    $594|    $670|
|------+--------+--------+--------+--------+--------+--------|
|1988  |  $2,412|    7.69|       4| $603.00|    $546|    $643|
|------+--------+--------+--------+--------+--------+--------|
|1987  |  $2,164|    6.90|       4| $541.00|    $484|    $595|
|------+--------+--------+--------+--------+--------+--------|
|1986  |  $1,922|    6.13|       4| $480.50|    $419|    $541|
|------+--------+--------+--------+--------+--------+--------|
|1985  |  $1,596|    5.09|       4| $399.00|    $337|    $448|
|------+--------+--------+--------+--------+--------+--------|
|1984  |  $1,528|    4.87|       4| $382.00|    $342|    $413|
|------+--------+--------+--------+--------+--------+--------|
|1983  |  $1,393|    4.44|       4| $348.25|    $299|    $384|
|------+--------+--------+--------+--------+--------+--------|
|1982  |  $1,252|    3.99|       4| $313.00|    $284|    $343|
|------+--------+--------+--------+--------+--------+--------|
|1981  |  $1,148|    3.66|       4| $287.00|    $247|    $323|
|------+--------+--------+--------+--------+--------+--------|
|1980  |  $1,030|    3.28|       4| $257.50|    $220|    $295|
|------+--------+--------+--------+--------+--------+--------|
|Total | $31,374|  100.00|      58| $540.93|    $220|    $998|
-----------------------------------------------------------------------
```

Creating Static HTML Output with SAS Tools

Before SAS 8, there were three choices for creating static HTML output:

- using the HTML `<pre>` and `</pre>` container tags to enclose results
- using a DATA step to insert HTML tags into SAS output
- using the HTML Formatting Tools, a series of SAS macros for generating Web pages

All of these are still available in Base SAS, but unless the only SAS release you have available at your site is one of these earlier versions, there is little reason to use the older tool set. ODS, the Output Delivery System introduced in SAS 7, can be used to produce a wide variety of output formats, including HTML and XML. Even though the earlier approaches are now superseded, it is instructive to examine how they work.

Displaying Preformatted Text

Display 3.2 displays the same information shown in Display 3.1, the output from the sample PROC TABULATE. This example was created by editing the SAS `.1st` file produced by the preceding program, saving it to the server as an HTML file (shown in Example 3.2), and displaying the result in a Web browser.

Example 3.2 Using HTML Preformatting

```
<html>
<head>
  <title>Retail Sales Table</title>
</head>
<body>
<h1 style="font-family: helvetica, tahoma, arial; color: blue">
  Retail Sales in Millions of $</h1>
<pre>
-------------------------------------------------------------
|Year  | Total  |Overall | Number |Average |Smallest|Largest |
|      | Sales  |Percent |of Sales|  Sale  |  Sale  |  Sale  |
|------+--------+--------+--------+--------+--------+--------|
|1994  |  $1,874|    5.97|       2| $937.00|   $876|    $998|
|------+--------+--------+--------+--------+--------+--------|
|1993  |  $3,578|   11.40|       4| $894.50|   $758|    $991|
|------+--------+--------+--------+--------+--------+--------|
|1992  |  $3,204|   10.21|       4| $801.00|   $692|    $889|
|------+--------+--------+--------+--------+--------+--------|
|1991  |  $2,947|    9.39|       4| $736.75|   $703|    $807|
|------+--------+--------+--------+--------+--------+--------|
|1990  |  $2,734|    8.71|       4| $683.50|   $606|    $749|
|------+--------+--------+--------+--------+--------+--------|
|1989  |  $2,592|    8.26|       4| $648.00|   $594|    $670|
|------+--------+--------+--------+--------+--------+--------|
|1988  |  $2,412|    7.69|       4| $603.00|   $546|    $643|
|------+--------+--------+--------+--------+--------+--------|
|1987  |  $2,164|    6.90|       4| $541.00|   $484|    $595|
|------+--------+--------+--------+--------+--------+--------|
|1986  |  $1,922|    6.13|       4| $480.50|   $419|    $541|
|------+--------+--------+--------+--------+--------+--------|
|1985  |  $1,596|    5.09|       4| $399.00|   $337|    $448|
|------+--------+--------+--------+--------+--------+--------|
```

```
|1984   |  $1,528|    4.87|         4| $382.00|   $342|   $413|
|------+--------+--------+--------+--------+--------+--------|
|1983   |  $1,393|    4.44|         4| $348.25|   $299|   $384|
|------+--------+--------+--------+--------+--------+--------|
|1982   |  $1,252|    3.99|         4| $313.00|   $284|   $343|
|------+--------+--------+--------+--------+--------+--------|
|1981   |  $1,148|    3.66|         4| $287.00|   $247|   $323|
|------+--------+--------+--------+--------+--------+--------|
|1980   |  $1,030|    3.28|         4| $257.50|   $220|   $295|
|------+--------+--------+--------+--------+--------+--------|
|Total  | $31,374|  100.00|        58| $540.93|   $220|   $998|
------------------------------------------------------------
</pre>
</body>
</html>
```

Display 3.2 HTML Preformatting Output

Retail Sales Table - Mozilla Firefox

File Edit View Go Bookmarks Tools Help

http://hygelac/BBU/example3-2.html

Retail Sales in Millions of $

```
------------------------------------------------------------
|Year   | Total  |Overall | Number |Average |Smallest|Largest |
|       | Sales  |Percent |of Sales| Sale   | Sale   | Sale   |
|------+--------+--------+--------+--------+--------+--------|
|1994   |  $1,874|    5.97|         2| $937.00|   $876|   $998|
|------+--------+--------+--------+--------+--------+--------|
|1993   |  $3,578|   11.40|         4| $894.50|   $758|   $991|
|------+--------+--------+--------+--------+--------+--------|
|1992   |  $3,204|   10.21|         4| $801.00|   $692|   $889|
|------+--------+--------+--------+--------+--------+--------|
|1991   |  $2,947|    9.39|         4| $736.75|   $703|   $807|
|------+--------+--------+--------+--------+--------+--------|
|1990   |  $2,734|    8.71|         4| $683.50|   $606|   $749|
|------+--------+--------+--------+--------+--------+--------|
|1989   |  $2,592|    8.26|         4| $648.00|   $594|   $670|
|------+--------+--------+--------+--------+--------+--------|
|1988   |  $2,412|    7.69|         4| $603.00|   $546|   $643|
|------+--------+--------+--------+--------+--------+--------|
|1987   |  $2,164|    6.90|         4| $541.00|   $484|   $595|
|------+--------+--------+--------+--------+--------+--------|
|1986   |  $1,922|    6.13|         4| $480.50|   $419|   $541|
|------+--------+--------+--------+--------+--------+--------|
|1985   |  $1,596|    5.09|         4| $399.00|   $337|   $448|
|------+--------+--------+--------+--------+--------+--------|
|1984   |  $1,528|    4.87|         4| $382.00|   $342|   $413|
|------+--------+--------+--------+--------+--------+--------|
|1983   |  $1,393|    4.44|         4| $348.25|   $299|   $384|
|------+--------+--------+--------+--------+--------+--------|
|1982   |  $1,252|    3.99|         4| $313.00|   $284|   $343|
|------+--------+--------+--------+--------+--------+--------|
|1981   |  $1,148|    3.66|         4| $287.00|   $247|   $323|
|------+--------+--------+--------+--------+--------+--------|
|1980   |  $1,030|    3.28|         4| $257.50|   $220|   $295|
|------+--------+--------+--------+--------+--------+--------|
|Total  | $31,374|  100.00|        58| $540.93|   $220|   $998|
------------------------------------------------------------
```

It is not necessary to change the SAS program in order to produce this output. Instead, open the **.1st** output file in your choice of text editor and then insert the following four lines above the title (HTML tags are shown in blue in the **.1st** file):

```
<html>
<head>
   <title>Retail Sales Table</title>
</head>
<body>
```

As described in Chapter 2, the effect of this code is to tell the browser this is an HTML page, and that the text "Retail Sales Table" should be displayed in the title bar of the browser window.

Next, format the title line so that it shows up as a heading, in a blue proportional font (the first specified font that is installed on the client system will be used to display the heading):

```
<h1 style="font-family: helvetica, tahoma, arial; color:
   blue">
   Retail Sales in Millions of $</h1>
```

The most important change is inserting the `<pre>` tag after the title but before the table; this indicates that the text is *preformatted*. Remember that HTML by default reduces two or more white space characters (such as a blank or a newline) to a single blank. If the table were not enclosed between the `<pre>` and `</pre>` tags, the browser would try to display the whole table as a single line, resulting in unintelligible gibberish. Finally, add the `</pre>`, `</body>` and `</html>` tags after the table to close off the page neatly.

There are several obvious drawbacks to this approach. First, you have to know at least some HTML in order to wrap the preformatted text in a reasonable context. More important, using preformatted character text discards all of the graphical tools available in HTML in favor of line printer mode. Finally, if you have a lot of text to format, this method can be very time consuming.

The advantage of SAS is that it saves time—surely there must be some way to automate this process. Of course, as any old-time SAS programmer will tell you, there is always a way to write any output format using PUT statements. (Old-time SAS programmers tend to be a little immodest.) In fact, it is possible to automate the production of HTML output by the creative use of that SAS standby, PROC PRINTTO.

DATA Step Programming for Web Output

The following example shows the output of a sample program for automatically formatting SAS procedure output into HTML. This is the kind of code that shows up in SAS Bowl questions at SUGI, but taken one step at a time, the effect is easy to understand.

Example 3.3 Using PUT Statements to Write HTML

```
filename temp 'Example 3-3.1st';
filename out  'Example 3-3.html';

*** redirect procedure output ***;
proc printto new file=temp;

*** create table ***;
proc tabulate data=SASHELP.RETAIL
   formchar='|----|+|---+=|-/\<>*';
```

```
        title 'Retail Sales In Millions Of $';
        class YEAR/descending;
        var SALES;

        table YEAR='' all='Total', SALES='' *
            (sum='Total Sales'*f=dollar8.
             pctsum='Overall Percent'*f=8.2
             n='Number of Sales'*f=8.
             mean='Average Sale'*f=dollar8.2
             min='Smallest Sale'*f=dollar8.
             max='Largest Sale'*f=dollar8.)/
             box='Year' rts=8;
  run;

  *** turn off redirection ****;
  proc printto;

  *** generate HTML output ***;
  data _null_;
    infile temp end=eof;
    input;
    file out;

    if (_n_ eq 1) then /* write header and title */
        put '<html>'/
            '<head>'/
            '<title>' _infile_ '</title>'/
            '<body>'/
            '<h1 style="color: blue">' _infile_ '</h1>'/
            '<pre>';
        else put _infile_;

    if (eof) then /* write closing tags */
        put '</pre>'/ '</body>'/' </html>';
  run;
```

At the beginning of the program two external files are referenced:

```
filename temp 'Example 3-3.lst';
filename out  'Example 3-3.html';
```

The SAS file reference `temp` is used to specify a temporary location for the SAS procedure output, whereas file reference `out` contains the resulting HTML page.

If your Web server is running Windows NT or Windows 2000, and the server is on a local Intranet, then it is possible for the SAS program to write the HTML output directly to the server directory where HTML files are stored. If the server is on a UNIX system or outside a firewall, this generally will not be possible.[2] In that case you will need to save your generated HTML temporarily and then use FTP or Secure FTP to transfer the Web pages to the server. (See Chapter 2 for more details.)

The PRINTTO procedure redirects all subsequent output to the temporary storage location. PROC TABULATE writes the table to the local output file. Then PROC PRINTTO is called again to close the output file (otherwise you cannot read from it in the same job step).

---

[2] If SAMBA is being used on the host, then SAS running on Windows can write output files to a UNIX operating system (see http://us1.samba.org/samba/what_is_samba.html).

The DATA step code reads in the temporary table from file `temp` and writes it to the output `out`. Using an INPUT statement with no other arguments has the effect of making the input buffer available for output using the PUT _INFILE_ statement.[3] This kind of programming endeared SAS to a generation of users, although it is somewhat out of fashion these days!

Three PUT statements are needed:

- one to write the header, write the title, and set up the preformatting
- one to copy the input buffer to output
- one to conclude the HTML page

The effect of all this is to write out HTML source code that is identical to the previous example, except that in this case DATA step programming has been substituted for manually using a text editor to insert the HTML tags. (The output of Example 3.3 is not shown because it looks exactly like that of Example 3.2).

This automated procedure runs very quickly, but it shares many of the disadvantages of the manual approach. You have to be familiar with PROC PRINTTO and the output format of the SAS procedures, plus you need to have a pretty good grasp of DATA step programming. If you want to take advantage of HTML formatting, you need to hard code the PUT statements for all of the tags by hand. In general, this approach is not very user friendly, is hard to maintain, and requires a lot of knowledge on the part of the user.

Macros for HTML Formatting

Release 6.12/6.09E of SAS introduced a set of automated macros that would take some of the burden of Web page creation off the user. Collectively, these macros are referred to as the HTML Formatting Tools; see http://support.sas.com/rnd/web/intrnet/format/.[4] As previously noted, these have been replaced by ODS for most purposes. It is still worthwhile to review how they work, however, both to help users who need to maintain and update legacy code as well as to understand the improvements in SAS Web technology.

The SAS macro language is a *metalanguage*, consisting of formal statements that generate SAS programming statements. Fairly large and elaborate SAS programs can be reduced greatly in size by using macro language to generate to program code. In addition, the code is likely to be more reliable, because the macro preprocessor is actually writing the statements. At the same time, using macros can be intimidating because of the highly compressed and symbolic notation.

The subject of SAS macro programming is far too large to cover here.[5] The first-time macro user should consult the extensive documentation available from SAS. To recap briefly, a reference to a name of the form `%MACNAME` causes the SAS statement processor to look for a macro called "MACNAME."

---

[3] If you are unfamiliar with the use in the sample program of the _N_ or _INFILE_ references, SAS supplies a number of automatic variables along with the other information in the DATA step. Among these are the current observation number (_N_) and the current input record (_INFILE_).

[4] Note that while specific URLs are included for the SAS online documentation, these may change in the future. The best advice is to go directly to http://support.sas.com, and then search for the most up-to-date information.

[5] See Art Carpenter, *Carpenter's Complete Guide to the SAS Macro Language*, 2nd ed. (Cary, NC: SAS Institute Inc., 2004).

SAS macros can take keyword parameters, for example:

```
%macname(parm1=value1,parm2=value2,...)
```

This statement will call the named macro and assign the corresponding value to each macro parameter in turn. Note that the SAS macro language has no variable typing; since it is a text substitution language, every variable is evaluated as a text string. Consequently, no quotes are necessary around character parameter values, and indeed may very well cause an error.

The super macro programmers at SAS have created a number of macros that can be used to create HTML output, as the following table illustrates:

Table 3.1 HTML Formatting Tools

HTML Formatting Tool	SAS Release
Base Formatting Tools:	
1. Output Formatter (%OUT2HTM) 2. Data Set Formatter (%DS2HTM) 3. Tabulate Formatter (%TAB2HTM)	Release 6.09 Enhanced Release 6.12 SAS 8 and later
SAS/GRAPH Software:	
4. GraphApplet HTML Generator without ActiveX support (%DS2GRAF)	Release 6.09 Enhanced Release 6.12
5. GraphApplet HTML Generator with ActiveX support (%DS2GRAF) 6. RangeView HTML Generator (%DS2CSF)	Release 6.09 Enhanced Release 6.12 SAS 8 and later
7. MetaView HTML Generator (%META2HTM)	SAS 8 and later
8. TreeView HTML Generator (%DS2TREE)	Release 8.2 and later
9. Constellation HTML Generator (%DS2CONST)	Release 9.1 and later

The six SAS/GRAPH macros are designed for specific functions that are beyond the scope of this discussion. If you are interested in using these, SAS has provided the details in the online reference cited previously. This section focuses on the first three, which are included with Base SAS and are the most generally useful.

%OUT2HTM – The Output Formatter

The HTML Output Formatter has been extensively revised and updated in SAS®9; see http://support.sas.com/rnd/web/intrnet/format/out/ for the details. As noted in the documentation, the resulting HTML file "doesn't just contain your output surrounded by `<pre>` tags; it contains your SAS output formatted using custom HTML tags…A predefined set of attributes, called properties, governs the formatting applied to the output." Using properties files, it is possible to customize the output by creating a property list. The *HTML Output Formatter Syntax Reference* at http://support.sas.com/rnd/web/intrnet/format/out/outsyn.html includes specific instructions for using this macro.

Example 3.4 shows how to use the output formatter to capture output from a SAS procedure. (PROC TABULATE has its own corresponding macro—%TAB2HTM. See the discussion later in this chapter.)

Example 3.4 HTML Output Formatter Macro

```
filename OUT   '../public_html/example3-4.html';

%OUT2HTM(capture=on);

*** create table ***;
proc means data=SASHELP.RETAIL
                  n mean min max
                  nonobs fw=8 maxdec=2;
        class YEAR/descending;
        var SALES;
title;
run;

%OUT2HTM(capture=off, htmlfref=OUT);
```

The Output Formatter macro is called twice, once before the procedure to turn on text capture, and once after it to generate the formatted page. In this way it is exactly analogous to the use of the PRINTTO procedure in the previous example. The macro is called initially with a single parameter, `capture=on`. After the MEANS procedure runs, the macro is called a second time, with three arguments:

```
%OUT2HTM(capture=off, htmlfref=out);
```

First, text capture is turned back off. The HTMLFREF= macro parameter specifies the SAS FILEREF of the resulting output, in this case `out`, which directs the generated HTML to `Example 3-4.html`. Alternatively, you can use the parameter

```
htmlfile= <external-filename>
```

in order to point to a specific file name. (Watch out that you don't use a FILEREF when you need a FILENAME, or vice versa!)

For the sake of convenience, this SAS program was run on the same system as the Web server. The statement

```
filename OUT   '../public_html/example3-4.html';
```

directs the output of the macro to the specified user directory. Of course if SAS is not available on the Web server, it would be possible to run this program locally and then transfer the output file to the server, as described in the preceding chapter. The resulting HTML page is shown in Display 3.3.

Display 3.3 %OUT2HTM Macro Output

Selecting View ▶ Source in the browser menu displays the following HTML generated by the macro.

Example 3.5 %OUT2HTM Macro HTML Source

```
<!DOCTYPE HTML PUBLIC "-//W3C//DTD HTML 3.2 Final//EN">
<HTML>

<HEAD>
  <META    NAME="GENERATOR"
        CONTENT="SAS Institute Inc. HTML Formatting Tools,
http://www.sas.com/">
  <TITLE></TITLE>
</HEAD>

<BODY>
```

```
<PRE><STRONG>The MEANS Procedure

Analysis Variable : SALES Retail sales in millions of $

    YEAR      N        Mean      Minimum     Maximum
------------------------------------------------------</STRONG></PRE>
<PRE>    1994      2      937.00      876.00      998.00

    1993      4      894.50      758.00      991.00

    1992      4      801.00      692.00      889.00

    1991      4      736.75      703.00      807.00

    1990      4      683.50      606.00      749.00

    1989      4      648.00      594.00      670.00

    1988      4      603.00      546.00      643.00

    1987      4      541.00      484.00      595.00

    1986      4      480.50      419.00      541.00

    1985      4      399.00      337.00      448.00

    1984      4      382.00      342.00      413.00

    1983      4      348.25      299.00      384.00

    1982      4      313.00      284.00      343.00

    1981      4      287.00      247.00      323.00

    1980      4      257.50      220.00      295.00</PRE>
<PRE><STRONG>
------------------------------------------------------

</STRONG></PRE>
<HR>

</BODY>

</HTML>
```

By today's standards the resulting code is pretty primitive. By default, the macro sets titles (not shown) as `<h3>` and the procedure heading as `<strong>`. The table itself is wrapped in `<pre>` and `</pre>` tags, just as in the previous two examples. You can of course pretty this up considerably; the designers built a great deal of flexibility into the macros, with a concomitant increase in complexity. For a complete list of all of the possible parameters for this macro, see the online *HTML Output Formatter Syntax Reference* at http://support.sas.com/rnd/web/intrnet/format/out/outsyn.html.

%DS2HTM – The Data Set Formatter

The HTML Data Set Formatter is quite a bit simpler than %OUT2HTM; it is intended simply to display the contents of a SAS data set. Example 3.6 illustrates the SAS code required to list a portion of the RETAIL data set using this formatter.

Example 3.6 HTML Data Set Formatter Macro

```
filename OUT '../public_html/example3-6.html';

title Sales Total by Month;

%ds2htm (
  data = SASHELP.RETAIL,
  where = YEAR gt 1990,
  var = YEAR MONTH SALES,
  htmlfref = out);
```

In this example, the macro is called with four parameters:

- *data* – the name of the SAS data set to be printed.

- *where* and *var* – code for the corresponding SAS program statements; note that the variable list is separated by spaces, not by commas, since these would then be read as macro parameters.

- *htmlfref* – the fileref of the output HTML file; use htmlfile to code the actual file name rather than the reference.

See the *HTML Data Set Formatter Syntax Reference* at http://support.sas.com/rnd/web/intrnet/format/ds/dssyn.html for a complete list of the possible macro options.

The resulting Web page is displayed below in Display 3.4.

Display 3.4 %DS2HTM Macro Output

Sales Total by Month

YEAR	MONTH	Retail sales in millions of $
1991	1	$703
1991	4	$709
1991	7	$728
1991	10	$807
1992	1	$692
1992	4	$797
1992	7	$826
1992	10	$889
1993	1	$758
1993	4	$909
1993	7	$920
1993	10	$991
1994	1	$876
1994	4	$998

Examining the generated HTML page source, note that SAS has created an HTML table statement (see Chapter 2 for a discussion of the `<table>` tag). The default formatting is again pretty simple, but as noted previously there are various ways to customize the output.

Example 3.7 %DS2HTM Macro HTML Source

```
<!DOCTYPE HTML PUBLIC "-//W3C//DTD HTML 3.2 Final//EN">
<HTML>
 <HEAD>
  <META     NAME="GENERATOR"
      CONTENT="SAS Institute Inc. HTML Formatting Tools,
http://www.sas.com/">
   <TITLE></TITLE>
</HEAD>

<BODY>

<PRE><H3>Sales Total by Month</H3></PRE>

<P>
<TABLE BORDER="1" WIDTH="100%" ALIGN="CENTER" CELLPADDING="1"
CELLSPACING="1">
  <TR>
    <TH ALIGN="CENTER" VALIGN="MIDDLE">YEAR</TH>
    <TH ALIGN="CENTER" VALIGN="MIDDLE">MONTH</TH>
    <TH ALIGN="CENTER" VALIGN="MIDDLE">Retail sales in millions of
  $</TH>
  </TR>
  <TR>
    <TD ALIGN="CENTER" VALIGN="MIDDLE">1991</TD>
    <TD ALIGN="CENTER" VALIGN="MIDDLE">1</TD>
    <TD ALIGN="CENTER" VALIGN="MIDDLE">      $703</TD>
  </TR>
  <TR>
    <TD ALIGN="CENTER" VALIGN="MIDDLE">1991</TD>
    <TD ALIGN="CENTER" VALIGN="MIDDLE">4</TD>
    <TD ALIGN="CENTER" VALIGN="MIDDLE">      $709</TD>
  </TR>
  <TR>
    <TD ALIGN="CENTER" VALIGN="MIDDLE">1991</TD>
    <TD ALIGN="CENTER" VALIGN="MIDDLE">7</TD>
    <TD ALIGN="CENTER" VALIGN="MIDDLE">      $728</TD>
  </TR>
  <TR>
    <TD ALIGN="CENTER" VALIGN="MIDDLE">1991</TD>
    <TD ALIGN="CENTER" VALIGN="MIDDLE">10</TD>
    <TD ALIGN="CENTER" VALIGN="MIDDLE">      $807</TD>
  </TR>
  <TR>
        [... repeated lines omitted ...]
  <TR>
    <TD ALIGN="CENTER" VALIGN="MIDDLE">1994</TD>
    <TD ALIGN="CENTER" VALIGN="MIDDLE">1</TD>
    <TD ALIGN="CENTER" VALIGN="MIDDLE">      $876</TD>
  </TR>
  <TR>
    <TD ALIGN="CENTER" VALIGN="MIDDLE">1994</TD>
```

```
      <TD ALIGN="CENTER" VALIGN="MIDDLE">4</TD>
      <TD ALIGN="CENTER" VALIGN="MIDDLE">        $998</TD>
   </TR>
</TABLE>
<P>

<HR>

</BODY>

</HTML>
```

%TAB2HTM – The Tabulate Formatter

The format macro for PROC TABULATE is basically just a version of the Output Formatter, with some special macro code to handle this procedure. Many of the parameter arguments are the same as for %OUT2HTM, but there are some additional table-specific ones (see the *HTML Tabulate Formatter Syntax Reference* at http://support.sas.com/rnd/web/intrnet/format/tab/tabsyn.html for more details).

Note the following message from the online documentation:

> **IMPORTANT**: Be sure that your program generates output only from PROC TABULATE while capturing is turned on. The HTML Tabulate Formatter will fail or produce unpredictable results if other types of output are also captured. To capture other types of output, use the HTML Output Formatter.

Example 3.8 illustrates the results of running the PROC TABULATE program with the %TAB2HTM macro.

Example 3.8 HTML Tabulate Formatter Macro

```
options noovp nodate nonumber nocenter ls=80
        formchar='82838485868788898a8b8c'x;

filename OUT  '../public_html/example3-8.html';

%TAB2HTM(capture=on);

*** create table ***;
proc tabulate data=SASHELP.RETAIL;

        title 'Retail Sales In Millions Of $';
        class YEAR/descending;
        var SALES;

        table YEAR='' all='Total', SALES='' *
              (sum='Total Sales'*f=dollar8.
               pctsum='Overall Percent'*f=8.2
               n='Number of Sales'*f=8.
               mean='Average Sale'*f=dollar8.2
               min='Smallest Sale'*f=dollar8.
               max='Largest Sale'*f=dollar8.)/
               box='Year' rts=8;
run;

%TAB2HTM(capture=off, htmlfref=OUT, center=Y);
```

You must include the FORMCHAR= option or no output will be displayed! As SAS *Usage Note V6-TAB2HTM-D742* explains:

> The HTML Formatting tool for TABULATE output, %TAB2HTM, requires that you use a special FORMCHARS string to enable the macro to correctly format the output. The FORMCHAR value is character set dependent as follows:

Character Set	Form Characters
ASCII	'82838485868788898A8B8C'x
EBCDIC	'B2B3B4B5B6B7B8B9BABBBC'x
DBCS	'02030405060708090A0B0C'x

Just as for %OUT2HTM, the Tabulate Formatter Macro is called twice, once before and once after the procedure. Even the parameter arguments for a minimal table are the same. The resulting Web page looks like Display 3.5:

Display 3.5 Tabulate Formatter Macro Output

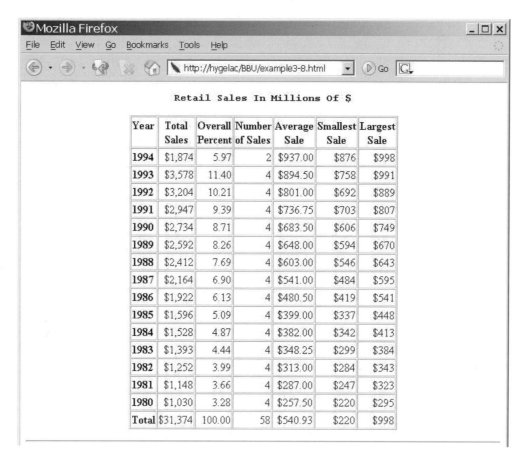

Looking at the resulting code, shown in Example 3.9, you can see that the output is structured as a `<table>`, as in the %DS2HTM macro, rather than using `<pre></pre>` tags like %OUT2HTM. That is probably why the documentation cautions you not to use one when you need the other, and not to mix them.

Example 3.9 Tabulate Formatter Macro HTML Source

```
<!DOCTYPE HTML PUBLIC "-//W3C//DTD HTML 3.2 Final//EN">
<HTML>

<HEAD>
  <META NAME="GENERATOR" CONTENT="SAS Institute Inc. HTML Formatting
Tools, http://www.sas.com/">
  <TITLE></TITLE>
</HEAD>

<BODY>

<CENTER>

<PRE><H3>Retail Sales In Millions Of $</H3></PRE>

<PRE><H3></H3></PRE>

<P>
<TABLE BORDER="1" WIDTH="0%" ALIGN="CENTER" CELLPADDING="1"
CELLSPACING="1">
  <TR>
 <TH ALIGN="LEFT" VALIGN="TOP" NOWRAP>Year</TH>
 <TH ALIGN="CENTER" VALIGN="BOTTOM" NOWRAP>Total<BR>Sales</TH>
 <TH ALIGN="CENTER" VALIGN="BOTTOM" NOWRAP>Overall<BR>Percent</TH>
 <TH ALIGN="CENTER" VALIGN="BOTTOM" NOWRAP>Number<BR>of Sales</TH>
 <TH ALIGN="CENTER" VALIGN="BOTTOM" NOWRAP>Average<BR>Sale</TH>
 <TH ALIGN="CENTER" VALIGN="BOTTOM" NOWRAP>Smallest<BR>Sale</TH>
 <TH ALIGN="CENTER" VALIGN="BOTTOM" NOWRAP>Largest<BR>Sale</TH>
  </TR>
  <TR>
 <TH ALIGN="LEFT" VALIGN="TOP" NOWRAP>1994</TH>
 <TD ALIGN="RIGHT" VALIGN="BOTTOM" NOWRAP>$1,874</TD>
 <TD ALIGN="RIGHT" VALIGN="BOTTOM" NOWRAP>5.97</TD>
 <TD ALIGN="RIGHT" VALIGN="BOTTOM" NOWRAP>2</TD>
 <TD ALIGN="RIGHT" VALIGN="BOTTOM" NOWRAP>$937.00</TD>
 <TD ALIGN="RIGHT" VALIGN="BOTTOM" NOWRAP>$876</TD>
 <TD ALIGN="RIGHT" VALIGN="BOTTOM" NOWRAP>$998</TD>
  </TR>
  <TR>
 <TH ALIGN="LEFT" VALIGN="TOP" NOWRAP>1993</TH>
 <TD ALIGN="RIGHT" VALIGN="BOTTOM" NOWRAP>$3,578</TD>
 <TD ALIGN="RIGHT" VALIGN="BOTTOM" NOWRAP>11.40</TD>
 <TD ALIGN="RIGHT" VALIGN="BOTTOM" NOWRAP>4</TD>
 <TD ALIGN="RIGHT" VALIGN="BOTTOM" NOWRAP>$894.50</TD>
 <TD ALIGN="RIGHT" VALIGN="BOTTOM" NOWRAP>$758</TD>
 <TD ALIGN="RIGHT" VALIGN="BOTTOM" NOWRAP>$991</TD>
  </TR>
        [... repeated lines omitted ...]
  <TR>
```

```
<TH ALIGN="LEFT" VALIGN="TOP" NOWRAP>1980</TH>
<TD ALIGN="RIGHT" VALIGN="BOTTOM" NOWRAP>$1,030</TD>
<TD ALIGN="RIGHT" VALIGN="BOTTOM" NOWRAP>3.28</TD>
<TD ALIGN="RIGHT" VALIGN="BOTTOM" NOWRAP>4</TD>
<TD ALIGN="RIGHT" VALIGN="BOTTOM" NOWRAP>$257.50</TD>
<TD ALIGN="RIGHT" VALIGN="BOTTOM" NOWRAP>$220</TD>
<TD ALIGN="RIGHT" VALIGN="BOTTOM" NOWRAP>$295</TD>
 </TR>
 <TR>
<TH ALIGN="LEFT" VALIGN="TOP" NOWRAP>Total</TH>
<TD ALIGN="RIGHT" VALIGN="BOTTOM" NOWRAP>$31,374</TD>
<TD ALIGN="RIGHT" VALIGN="BOTTOM" NOWRAP>100.00</TD>
<TD ALIGN="RIGHT" VALIGN="BOTTOM" NOWRAP>58</TD>
<TD ALIGN="RIGHT" VALIGN="BOTTOM" NOWRAP>$540.93</TD>
<TD ALIGN="RIGHT" VALIGN="BOTTOM" NOWRAP>$220</TD>
<TD ALIGN="RIGHT" VALIGN="BOTTOM" NOWRAP>$998</TD>
 </TR>
</TABLE>

<P>

<P>
<HR><P>

</CENTER>

</BODY>

</HTML>
```

Note that it is possible to generate dynamic Web pages using the SAS Web Publishing macros, using the options HTMLFREF=_WEBOUT and RUNMODE=S. Since these require the use of the Application Dispatcher in SAS/IntrNet, discussion of these options will be deferred until Chapter 6, "SAS/IntrNet: the Application Dispatcher," where this topic is introduced.

Using the Output Delivery System to Create HTML

After all of this elaborate discussion, it seems almost anticlimactic that to create static HTML in SAS 8 and later, all you need to do is check the **Create HTML** box on the **Results** tab of the **Preferences** dialog box in the SAS Display Manager. You even get a choice of numerous predefined styles, displayed in the **Style** list, as illustrated in Display 3.6.

Display 3.6 Create HTML in the Display Manager Preferences

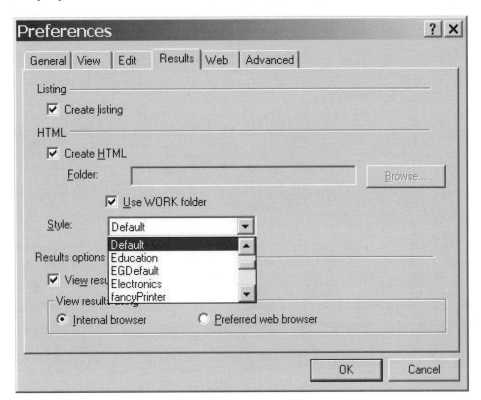

New HTML pages are created by default in the SAS Temporary Files folder as a document called **sashtm.htm**. If you have the **View results as generated** box checked, they are also displayed in the Results window. From this latter window, you can go to the **File** menu and save the generated page to a selected directory. (Incidentally, this feature is not only available on Windows; it works the same way in the UNIX and Linux versions of SAS). You can uncheck the **Create listing** box to suppress the normal listing to the output window.

Display 3.7 illustrates the page generated by running the PROC TABULATE example from the SAS Display Manager with the **Create HTML** preference checked and the default style.

Display 3.7 Using the SAS Display Manager to Create HTML

SAS Output - Mozilla Firefox

File   Edit   View   Go   Bookmarks   Tools   Help

http://hygelac/BBU/example3-10.html   Go

Retail Sales In Millions Of $

Year	Total Sales	Overall Percent	Number of Sales	Average Sale	Smallest Sale	Largest Sale
1994	$1,874	5.97	2	$937.00	$876	$998
1993	$3,578	11.40	4	$894.50	$758	$991
1992	$3,204	10.21	4	$801.00	$692	$889
1991	$2,947	9.39	4	$736.75	$703	$807
1990	$2,734	8.71	4	$683.50	$606	$749
1989	$2,592	8.26	4	$648.00	$594	$670
1988	$2,412	7.69	4	$603.00	$546	$643
1987	$2,164	6.90	4	$541.00	$484	$595
1986	$1,922	6.13	4	$480.50	$419	$541
1985	$1,596	5.09	4	$399.00	$337	$448
1984	$1,528	4.87	4	$382.00	$342	$413
1983	$1,393	4.44	4	$348.25	$299	$384
1982	$1,252	3.99	4	$313.00	$284	$343
1981	$1,148	3.66	4	$287.00	$247	$323
1980	$1,030	3.28	4	$257.50	$220	$295
Total	$31,374	100.00	58	$540.93	$220	$998

The underlying source code for this page is somewhat easier to understand after some Web programming practice. For the moment, it is sufficient to note that there is a whole lot more stuff in this page than we have seen until now. (The folks at SAS have obviously been keeping busy.) You should be able to see that this page does not use the `<pre>` and `</pre>` tags like %OUT2HTM. Instead, an HTML table is created, with a number of formatting instructions determined by the CSS style selected.

The header for the page also includes some JavaScript functions, enclosed in `<SCRIPT></SCRIPT>` tags, that are called when the page is loaded and unloaded. In this case, since no special formatting was requested, the function bodies are empty, except to figure out whether the client browser is on Windows or not. (As will be discussed in Chapter 5, "Web Applications Programming," JavaScript usually executes on the client system, so the client OS matters.)

Note that the HTML generated does not conform to the XHTML standard, but it does work in most Web browsers. (See the discussion of the ODS MARKUP statement in the section later in this chapter, "Using the ODS Markup Statement to Create XHTML," for information about how to create XHTML using SAS.) Note that an internal CSS file is created by default. The following example shows only the first of these generated style tags—`.ContentTitle`. The remaining ones are omitted from the listing for reasons of space.

Example 3.10 SAS Display Manager HTML Source

```
<!DOCTYPE HTML PUBLIC "-//W3C//DTD HTML 4.0 Transitional//EN">
<HTML><HEAD><TITLE>SAS Output</TITLE>
<META content="MSHTML 6.00.2900.2802" name=GENERATOR sasversion="9.1">
<META http-equiv=Content-type content="text/html; charset=windows-
1252">
<STYLE type=text/css>.ContentTitle {
   [... a whole bunch of CSS styles omitted ...]
   }
</STYLE>
<SCRIPT language=javascript type=text/javascript>
<!--
function startup(){

}
function shutdown(){

}

//-->
</SCRIPT>
<NOSCRIPT></NOSCRIPT></HEAD>
<BODY class=Body onload=startup() onunload=shutdown()>
<SCRIPT language=javascript type=text/javascript>
<!--
var _info = navigator.userAgent
var _ie = (_info.indexOf("MSIE") > 0
         && _info.indexOf("Win") > 0
         && _info.indexOf("Windows 3.1") < 0);

//-->
</SCRIPT>
<NOSCRIPT></NOSCRIPT>
<DIV class=branch><A name=IDX></A>
<TABLE class=SysTitleAndFooterContainer cellSpacing=1 cellPadding=1
rules=none width="100%" summary="Page Layout" border=0 frame=void>
```

```
      <TBODY>
      <TR>
         <TD class="l SystemTitle">Retail Sales In Millions Of
         $</TD></TR></TBODY></TABLE><BR>
<DIV>
<DIV align=left>
<TABLE class=Table borderColor=#000000 cellSpacing=1 cellPadding=7
rules=groups summary="Procedure Tabulate: Table 1" border=1 frame=box>
      <COLGROUP>
      <COL></COLGROUP>
      <COLGROUP>
      <COL>
      <COL>
      <COL>
      <COL>
      <COL>
      <COL></COLGROUP>
      <THEAD>
      <TR>
         <TH class="c m Header" scope=col>Year</TH>
         <TH class="c Header" scope=col>Total Sales</TH>
         <TH class="c Header" scope=col>Overall Percent</TH>
         <TH class="c Header" scope=col>Number of Sales</TH>
         <TH class="c Header" scope=col>Average Sale</TH>
         <TH class="c Header" scope=col>Smallest Sale</TH>
         <TH class="c Header" scope=col>Largest Sale</TH></TR></THEAD>
      <TBODY>
      <TR>
         <TH class="l t RowHeader" scope=row>1994</TH>
         <TD class="r b Data">$1,874</TD>
         <TD class="r b Data">5.97</TD>
         <TD class="r b Data">2</TD>
         <TD class="r b Data">$937.00</TD>
         <TD class="r b Data">$876</TD>
         <TD class="r b Data">$998</TD></TR>
            [... repeated lines omitted ...]
      <TR>
         <TH class="l t RowHeader" scope=row>Total</TH>
         <TD class="r b Data">$31,374</TD>
         <TD class="r b Data">100.00</TD>
         <TD class="r b Data">58</TD>
         <TD class="r b Data">$540.93</TD>
         <TD class="r b Data">$220</TD>
         <TD class="r b Data">$998</TD></TR>
</TBODY></TABLE></DIV></DIV><BR></DIV></BODY></HTML>
```

But, as all good SAS programmers know, you cannot always run everything from the Display Manager window (although this would certainly be the method of choice for developing and debugging static Web output). It is not difficult to learn how to use the available ODS commands in batch mode to create Web pages of almost any format.

The ODS HTML Statement in SAS®9

Up to and including Release 6.12 of SAS, pretty much every procedure had its own routines for handling printed output. Many of these dated back to the original versions of SAS developed in the early 1970s, when the only output devices available were line printers and TTY terminals. As each new release of SAS appeared, the basic output formatting code was rewritten, but every procedure still managed its own layout. SAS users learned to use PROC PRINTTO, DDE triplets, and other exotic solutions in order to impose their own formats on output, or to select specific portions of the results for further processing.

One of the prime directives of object-oriented software development is "Thou shall not mix Data with Presentation." Starting with SAS 7, the Output Delivery System was introduced to manage all the SAS procedure output in a consistent way. One or more *output objects* are created by each DATA step or procedure; these output objects each contain two basic components.[6] The *data* component includes the raw data values that comprise the results of the procedure or the contents of the Program Data Vector (PDV), while the *table* template (also called a table definition) describes how the data should be formatted. SAS has supplied a number of standard templates. You can modify these and even create your own using PROC TEMPLATE.[7]

As the online SAS System Help points out:

> ODS removes responsibility for formatting output from individual procedures and from the DATA step. The procedure or DATA step supplies raw data and the name of the table definition that contains the formatting instructions, and ODS formats the output. Because formatting is now centralized in ODS, the addition of a new ODS destination does not affect any procedures or the DATA step. As future destinations are added to ODS, they will automatically become available to all procedures that support ODS and to the DATA step.

The **Create HTML** check box selection described previously in fact just uses ODS. The drop-down menu shows you the list of the production style templates stored in the SASHELP.TMPLMST item store, so that you can pick the presentation style of the resulting output file. You can easily create your own HTML in batch mode, using a few simple ODS statements that will accomplish the same effect. Example 3.11 shows how to do it; the HTML output and page source code are identical to that shown in Display 3.6 and Display 3.7.

Example 3.11 Using the ODS HTML Statement

```
filename OUT "Example 3-11.html";

ods listing close;
ods html body=OUT;

proc tabulate data=SASHELP.RETAIL;
  title "Retail Sales In Millions of $";
  class YEAR/descending;
  var SALES;
  table YEAR="" all="Total", SALES="" *
  (sum="Total Sales"*f=dollar8.
  pctsum="Overall Percent"*f=8.2
```

---

[6] This discussion borrows heavily from Chris Olinger, "ODS for Dummies" (Paper 25-64, Proceedings of the Twenty-fifth Annual SAS Users Group International Conference. Cary, NC: SAS Institute Inc., 2000).

[7] C++ users should note that SAS uses the term "standard template" in a way that is quite different from what they may be used to. For more information about working with SAS templates, see the *PROC TEMPLATE FAQ and Concepts* at http://support.sas.com/rnd/base/topics/templateFAQ/Template.html.

```
    n="Number of Sales"*f=8.
    mean="Average Sale"*f=dollar8.2
    min="Smallest Sale"*f=dollar8.
    max="Largest Sale"*f=dollar8.)/
    box="Year" rts=8;
run;


ods html close;
ods listing;
```

ODS statements open and close specified destinations. In this example, ODS HTML and ODS LISTING statements are used to manage the output process. It is still necessary to call ODS twice, once before and once after the procedure. But with the earlier approach, the first invocation of the macro turned on a routine to capture the printer output from the procedure. The second call to the macro then added HTML tags and copied the result to the specified output location. In this example, however, the first ODS HTML statement opens the output location, and from that point on, all procedure and data step output are sent to the file specified in the BODY= option until it is explicitly closed. The two ODS LISTING statements are used to suppress the standard output listing while the ODS process is underway, and to restart it afterwards.

From the point of view of the user, ODS appears to work in a way that is very similar to the %OUT2HTM macro. In fact, what is under the hood is completely different, as the generated page source code illustrates. Instead of using preformatting, ODS dynamically generates the output in the specified style. But that's just the beginning of what you can do with ODS. Although this book is not an ODS tutorial, the Output Delivery System is so much a part of Web programming that it is important to point out a few of the most notable features.

Creating Multiple Pages with a Single Program

One of the most important improvements in ODS of previous approaches is its ability to manage multiple output objects. The following example shows the code to create a single new HTML page for each procedure.

Example 3.12 Creating Multiple Web Pages

```
ods listing close;
ods html body="../public_html/example3-12.html" newfile=proc;

***** Step #1 ****;
proc print data=SASHELP.RETAIL;
    title "1994 Sales Total by Month";
    where YEAR gt 1990;
    var MONTH SALES;
    id YEAR;
run;

***** Step #2 ****;
proc means data=SASHELP.RETAIL
    n mean min max
    nonobs fw=8 maxdec=2;
    title 'Retail Sales In Millions Of $';
    class YEAR/descending;
    var SALES;
run;
```

```
***** Step #3 ****;
proc tabulate data=SASHELP.RETAIL;

    class YEAR/descending;
    var SALES;
    table YEAR='' all='Total', SALES='' *
        (sum='Total Sales'*f=dollar8.
         pctsum='Overall Percent'*f=8.2
         n='Number of Sales'*f=8.
         mean='Average Sale'*f=dollar8.2
         min='Smallest Sale'*f=dollar8.
         max='Largest Sale'*f=dollar8.)/
    box='Year' rts=8;
run;

ods html close;
ods listing;
```

The ODS HTML statement in this example differs from the one in Example 3.11 in that the ODS NEWFILE= option is used to indicate that a new HTML file should be started for each procedure. ODS automatically increments the filename for each new procedure. It is useful to recognize that ODS increments the *rightmost* number that it finds. For example, if the above example instead had used the statement `ods html body="example3-12_f1.html"` then the subsequent files would have been named as follows:

```
example3-12_f1.html
example3-12_f2.html
example3-12_f3.html
```

For this reason, the file name must be specified in the BODY= option; a file reference cannot be used.

Display 3.8 shows a portion of the resulting SAS log file. As the SAS Notes indicate, three output HTML files are created, one for each of the three procedures.

Display 3.8 Creating Multiple Web Pages with a Single Program

```
1          options noovp nodate nonumber nocenter ls=80;
2
3          ods listing close;
4          ods html body="../public_html/example3-12.html"
newfile=proc;

NOTE: Writing HTML Body file: ../public_html/example3-12.html
5
6          ***** Step #1 ****;
7          proc print data=SASHELP.RETAIL;
8             title "1994 Sales Total by Month";
9             where YEAR gt 1990;
10            var MONTH SALES;
11            id YEAR;
12         run;

NOTE: There were 14 observations read from the data set
SASHELP.RETAIL.
      WHERE YEAR>1990;
```

```
NOTE: PROCEDURE PRINT used (Total process time):
      real time           0.08 seconds
      cpu time            0.08 seconds

13
14          ***** Step #2 ****;
15          proc means data=SASHELP.RETAIL
16                  n mean min max
17                  nonobs fw=8 maxdec=2;
18              title 'Retail Sales In Millions Of $';
19              class YEAR/descending;
20              var SALES;
21          run;

NOTE: Writing HTML Body file: ../public_html/example3-13.html

NOTE: There were 58 observations read from the data set
SASHELP.RETAIL.
NOTE: PROCEDURE MEANS used (Total process time):
      real time           0.18 seconds
      cpu time            0.18 seconds

22
23          ***** Step #3 ****;
24          proc tabulate data=SASHELP.RETAIL;
25
26            class YEAR/descending;
27            var SALES;
28            table YEAR='' all='Total', SALES='' *
29                  (sum='Total Sales'*f=dollar8.
30                   pctsum='Overall Percent'*f=8.2
31                   n='Number of Sales'*f=8.
32                   mean='Average Sale'*f=dollar8.2
33                   min='Smallest Sale'*f=dollar8.
34                   max='Largest Sale'*f=dollar8.)/
35            box='Year' rts=8;
36          run;

NOTE: Writing HTML Body file: ../public_html/example3-14.html

NOTE: There were 58 observations read from the data set
SASHELP.RETAIL.
NOTE: PROCEDURE TABULATE used (Total process time):
      real time           0.23 seconds
      cpu time            0.23 seconds

37
38          ods html close;
39          ods listing;
40

NOTE: SAS Institute Inc., SAS Campus Drive, Cary, NC USA 27513-2414
NOTE: The SAS System used:
      real time           0.93 seconds
      cpu time            0.30 seconds
```

The NEWFILE= option can take on one of five values:

- NONE is the default, and writes all output to the current body file.
- PROC writes one HTML file per procedure, as shown in the example.
- BYGROUP writes one HTML file for each BY group value.
- OUTPUT writes one HTML file for each output object. Some procedures, like PROC REG, produce more than one output object.
- PAGE writes a new file for each page of output.

Based on the value of the option selected, the user can split complex SAS printed output across any number of HTML pages. Because large pages can take a long time to load, this can be a useful trick.

Creating a Table of Contents for Procedure Output

Another neat thing you can do with ODS is to automatically generate a table of contents. Example 3.13 shows how the Output Delivery System can be used to generate a frame page linking multiple HTML documents.

Example 3.13 Creating Frames with ODS

```
ods listing close;
ods html path="../public_html" (url="http://hygelac/BBU/")
        body = "body3-13.html"
        contents = "contents3-13.html"
        frame = "frame3-13.html";

***** Step #1 ****;
proc print data=SASHELP.RETAIL;
   title "1994 Sales Total by Month";
   where YEAR gt 1990;
   var MONTH SALES;
   id YEAR;
run;

***** Step #2 ****;
proc means data=SASHELP.RETAIL
      n mean min max
       nonobs fw=8 maxdec=2;
     title 'Retail Sales In Millions Of $';
     class YEAR/descending;
     var SALES;
run;

***** Step #3 ****;
proc tabulate data=SASHELP.RETAIL;

   class YEAR/descending;
   var SALES;
   table YEAR='' all='Total', SALES='' *
       (sum='Total Sales'*f=dollar8.
        pctsum='Overall Percent'*f=8.2
        n='Number of Sales'*f=8.
        mean='Average Sale'*f=dollar8.2
```

```
         min='Smallest Sale'*f=dollar8.
         max='Largest Sale'*f=dollar8.)/
  box='Year' rts=8;
run;

ods html close;
ods listing;
```

Opening the frame URL displays all three pages, as illustrated in Display 3.9. Clicking on a link in the left-hand window brings up the corresponding portion of the body. In addition to providing an absolute URL, it is also possible to specify `URL="NONE"`; the resulting HTML will include a anchor tag with a relative URL of the form `<a href="filename.ext">`.

This example shows a very simplistic page style. By using some simple SAS code, it is possible to format the table of contents so that it will have almost any desired appearance. There are many user-written papers about this topic; see the references at the end of this chapter for more information.

Display 3.9 Using ODS to Display Multiple Pages in a Frame

In this example, the ODS statement has four arguments:

- `path` – directory location for the three output files (see below for an explanation of the URL option)
- `body` – body3-13.html
- `contents` – contents3-13.html
- `frame` – frame3-13.html

The body file contains all the generated HTML pages; the contents file includes links to the various parts of the body; and the frame file pulls it all together. Note that the hyperlink in this case must be to the frame file, since this file includes anchor tags to the content and body pages.

These three files are written to the local directory specified by the ODS path option:

```
path="..\public_html"
```

that is, the current user's public Web directory on the server. However, the anchor tags in the contents page use the URL option url=<u>http://hygelac/BBU/</u>. Without the URL option, the pages would not be operable if moved to a different server, since in that case the anchor links in the frame would point to the wrong platform.

Listing the Contents of a Data Set

It would be remiss not to demonstrate how easy it is to use ODS with the DATA step to list the contents of a SAS data set. The trick is a new SAS automatic variable called _ODS_. Example 3.14 shows how to create a DATA step listing with ODS.

Example 3.14 ODS HTML Data Set Listing

```
filename OUT  "../public_html/example3-14.html";
title "Sales Total by Month 1991-1994";

ods listing close;
ods html body=OUT;

data _null_;
  set SASHELP.RETAIL;
  where (YEAR gt 1990);
  file print ods=(variables=(YEAR MONTH SALES));
  put _ods_;
run;

ods html close;
ods listing;
```

Remember that a SAS data set is essentially a table, where the rows are observations and the columns are variables. In this example, the WHERE statement selects a subset of the observations, while the VARIABLES= sub-option on the FILE PRINT ODS= option controls which variables will be listed and in what order. Watch out for the parentheses! They have to be included exactly as shown; the sub-option is in one set of brackets, whereas the variable list has to be enclosed in an inner set.

The statement put _ods_ sends the requested data items to HTML. The result is a nicely formatted listing that is a handy alternative to PROC PRINT. The output looks like this (compare to Display 3.5, where the same data is displayed using the %DS2HTM formatting macro):

Display 3.10 ODS HTML Data Set Listing

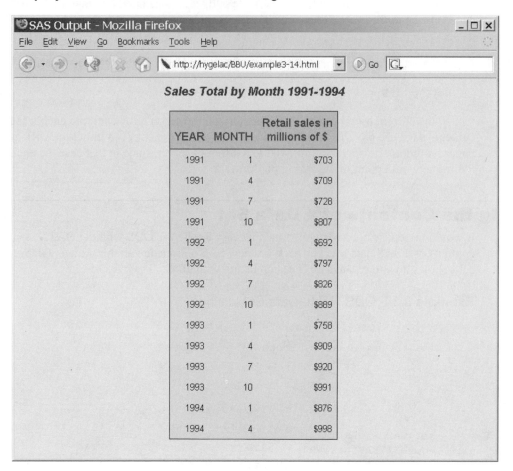

Obviously all these tools could be combined. For example, you could easily generate multiple SAS HTML files, including the output from a DATA step, with a table of contents pointing to all of them.

Using Styles to Format ODS Output

One last ODS HTML option needs to be discussed before we can logically leave this topic. Unfortunately, it is a huge one, and we can only nibble around the edges of what is available. This is the STYLE= option, used to specify a style template (also known as a style definition) for HTML. All of the examples we have seen so far have used the default style, but there are lots of other options, plus a considerable amount of customization available.

The SAS setup routine in SAS 8 and later automatically creates a library file called an *item store*. This file stores the standard ODS templates; by default, it is called SASHELP.TMPLMST. The item store actually holds several different kinds of templates, the most important of which are *table* and *style*. The table templates are used by SAS to provide basic table layouts. The style templates, on the other hand, describe the presentation of the information, such as color, font, and alignment.

You can also create your own templates; by default these are stored in SASUSER.TEMPLAT. Be careful though; if SAS is reinstalled, the SASUSER library is reinitialized and all your templates will go away. Keep a backup, just in case. (Note that you cannot move sashelp.tmplmst or sasuser.templat across platforms. You should keep the SAS code that created the template, and not just make a backup copy of the item store.)

If you are interested in the details of this process, select **View ▶ Results** from the main window in SAS Display Manager. From the **Results** window, again select **View**. The menu should have changed so that the first item is **Templates**. Click **Templates** and then expand SASHELP.TMPLMST by clicking on the + sign. (You can also get this information by typing odstemplate in the task bar or command line.)

You should see something like the following window:

Display 3.11 SAS Template Library

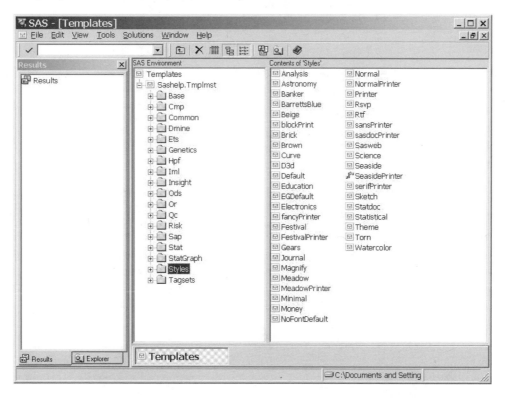

Clicking on one of the styles brings up the SAS code for the corresponding TEMPLATE procedure. For example, the **Default** style template starts out like this (and goes on at some length):

Example 3.15 Default Style Template

```
proc template;

  define style Styles.Default;

    style fonts
      "Fonts used in the default style" /
      'TitleFont2' = ("Arial, Helvetica, sans-serif",4,Bold Italic)
      'TitleFont' = ("Arial, Helvetica, sans-serif",5,Bold Italic)
      'StrongFont' = ("Arial, Helvetica, sans-serif",4,Bold)
      'EmphasisFont' = ("Arial, Helvetica, sans-serif",3,Italic)
      'FixedEmphasisFont' = ("Courier New, Courier,
        monospace",2,Italic)
      'FixedStrongFont' = ("Courier New, Courier,
        monospace",2,Bold)
      'FixedHeadingFont' = ("Courier New, Courier, monospace",2)
      'BatchFixedFont' = ("SAS Monospace, Courier New, Courier,
        monospace",2)
      'FixedFont' = ("Courier",2)
      'headingEmphasisFont' = ("Arial, Helvetica, sans-
        serif",4,Bold Italic)
      'headingFont' = ("Arial, Helvetica, sans-serif",4,Bold)
      'docFont' = ("Arial, Helvetica, sans-serif",3);
    [ ... many more styles omitted ... ]
  end;
run;
```

Once you get into it, you can use the TEMPLATE procedure to modify existing styles and even create your own. To specify a style on the ODS HTML statement, just use the STYLE= option with one of the style templates. For example:

Example 3.16 ODS Sasweb Style Data Set Listing

```
filename OUT  "../public_html/example3-16.html";
title "Sales Total by Month 1991-1994";

ods listing close;
ods html body=OUT style=sasweb;

data _null_;
  set SASHELP.RETAIL;
  where (YEAR gt 1990);
  file print ods=(variables=(YEAR MONTH SALES));
  put _ods_;
run;

ods html close;
ods listing;
```

The **sasweb** style is a nice shade of blue (compare to the output in Display 3.10):

Display 3.12 ODS HTML Sasweb Style

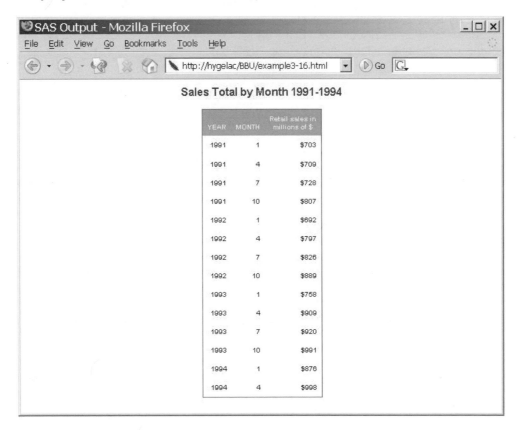

There are a great many other formatting options available with the ODS. Some of the most useful include the following:

- The STYLESHEET option by itself (no = sign or file specification) inserts an internal CSS `<style>` section into the created HTML page; see the preceding chapter for a discussion of cascading style sheets and image files).

- STYLESHEET=*newcssfilename.css* creates an external CSS file and puts an HTML `<link>` tag in the created HTML page that points to the newly created CSS file.

- STYLESHEET=(URL=*existingcssfile.css*) puts an HTML `<link>` tag in the created HTML page that points to an already existing `css` file which should contain the HTML and CLASS selectors expected by ODS, as well as any corporate CLASS or other selectors that are required.

The first choice, the STYLESHEET option by itself, was necessary in SAS 8 to generate an inline `<style>` section; in SAS®9 this is the default. In other words, in SAS®9 with ODS HTML you get an inline `<style>` section without having to specify the option.

PROC TEMPLATE is available to modify existing styles or create your own from scratch; this procedure can also be used to create custom TABLE templates to be used with SAS procedures instead of the default TABLE templates. ODS can create a GIF file of SAS/GRAPH output and an

HTML file with an `<img>` tag to display the graph. Also, SAS now allows the use of SAS/GRAPH style format controls in the TITLE, FOOTNOTE, and LABEL statements.

All these topics (and more) await you when you begin to explore the resources SAS has provided for formatting ODS output; see http://support.sas.com/rnd/base/index-ods-resources.html for more information on the resources available for the Output Delivery System. There is also a site for customer-contributed tagsets (including XHTML) at http://support.sas.com/rnd/base/topics/odsmarkup/customer.html.

Using the ODS Markup Statement to Create XHTML

All of the discussion about creating Web pages in this chapter thus far has applied mostly to HTML 4.01, but HTML is only one kind of markup language. As noted earlier, the existing specification has a major drawback in that the HTML standard includes both data and the instructions for displaying that information. Mixing the data and the presentation in this way is considered a no-no in object-oriented programming circles, because it makes it difficult to change one and not the other.

The solution (proposed in February 1998) is *Extensible Markup Language* (XML). XML eliminates any formatting instructions in favor of separate stylesheets. (See the SUGI papers referenced at the end of this chapter for more detail on how to create XML using SAS.) XML immediately became the medium of choice in the industry for cross-platform data transfer, but it has not replaced HTML for data presentation. This is probably due to two reasons: first, the large body of existing HTML code, and second, the additional complication of learning and maintaining the new XML format.

In response to this singular lack of enthusiasm, in January 2000 the W3C issued a new recommendation, "XHTML 1.0: The Extensible HyperText Markup Language: A Reformulation of HTML 4 in XML 1.0" (http://www.w3.org/TR/xhtml1/). This new standard was subsequently updated to XHTML 1.1 in May 2001 as "XHTML 1.1 – Module-based XHTML" (http://www.w3.org/TR/2001/REC-xhtml11-20010531). As the recommendation notes, XHTML is intended to replace HTML, which is now deprecated by the W3C. While HTML is still supported, it will go away at some point and be replaced by XHTML, which is an XML-based hypertext markup language. (See the section on "HTML vs. XHTML" in Chapter 2 for more information about XHTML.)

The folks at SAS have recognized that the application technology necessary to write HTML is pretty much the same as that needed to write XML, XSL, CSV (comma-separated values), DTD (XML Document Type Definition), and CSS. The ODS MARKUP statement was introduced in SAS 8.2 in order to allow the creation of files in a variety of markup languages. As explained in the article "Using ODS to Export Markup Languages" (http://support.sas.com/rnd/base/topics/odsmarkup/), the syntax for ODS MARKUP is basically the same as ODS HTML, except that ODS MARKUP has the TAGSET= option. In future releases of SAS, it is to be expected that ODS HTML will be deprecated and ODS MARKUP will be the method of choice.

The TAGSET= option is used to specify the type of output required. In SAS 8.2, over 20 different destinations are supported. The Templates window (see Display 3.11) can also be used to list the available tagsets.

As Example 3.17 illustrates, if you know how to use the ODS HTML statement, you can use ODS MARKUP to create XHTML.

Example 3.17 ODS Markup: XHTML

```
filename OUT "../public_html/example3-17.html";

ods listing close;
ods markup body=OUT tagset=XHTML;

proc tabulate data=SASHELP.RETAIL;
   title "Retail Sales In Millions Of $";
   class YEAR/descending;
   var SALES;
   table YEAR="" all="Total", SALES="" *
       (sum="Total Sales"*f=dollar8.
        pctsum="Overall Percent"*f=8.2
        n="Number of Sales"*f=8.
        mean="Average Sale"*f=dollar8.2
        min="Smallest Sale"*f=dollar8.
        max="Largest Sale"*f=dollar8.)/
        box="Year" rts=8;

run;

ods markup close;
ods listing;
```

Using a customized template (see the next section), it is possible to create the XHTML Web page shown in Display 3.13.

Display 3.13 ODS Markup: XHTML Output

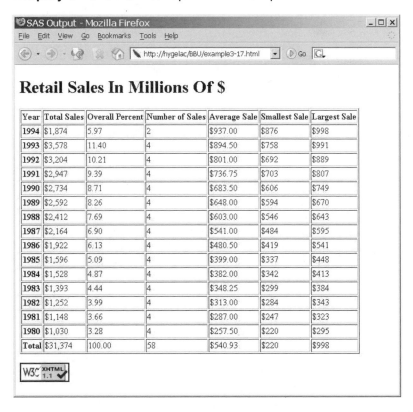

The really interesting part of this example is the page source code, as shown in Example 3.18. The following section explains how to create XHTML (or any other markup) using a custom tagset and ODS.

Example 3.18 XHTML Source

```
<?xml version="1.0" encoding="UTF-8"?>
<!DOCTYPE html PUBLIC "-//W3C//DTD XHTML 1.1//EN"
   "http://www.w3.org/TR/xhtml1/DTD/xhtml11.dtd">
<html xmlns="http://www.w3.org/1999/xhtml">
   <head>
      <title>SAS Output</title>
   </head>
<body>
   <h1>Retail Sales In Millions Of $</h1>
   <table border="1">
      <thead>
         <tr>
            <th>Year</th><th>Total Sales</th>
            <th>Overall Percent</th>
            <th>Number of Sales</th>
            <th>Average Sale</th><th>Smallest Sale</th>
            <th>Largest Sale</th>
         </tr>
      </thead>
      <tbody>
         <tr>
            <th>   1994</th>
            <td>  $1,874</td>
            <td>    5.97</td>
            <td>       2</td>
            <td> $937.00</td>
            <td>    $876</td>
            <td>    $998</td>
         </tr>
      [... repeated lines omitted ...]
         <tr>
            <th>   1980</th>
            <td>  $1,030</td>
            <td>    3.28</td>
            <td>       4</td>
            <td> $257.50</td>
            <td>    $220</td>
            <td>    $295</td>
         </tr>
         <tr>
            <th>Total</th>
            <td> $31,374</td>
            <td>  100.00</td>
            <td>      58</td>
            <td> $540.93</td>
            <td>    $220</td>
            <td>    $998</td>
         </tr>
      </tbody>
   </table>
   <p>
   <a href="http://validator.w3.org/check/referer">
```

```
    <img src="http://www.w3.org/Icons/valid-xhtml11"
      alt="Valid XHTML 1.1!" height="31" width="88" /></a>
    </p>
    </body>
</html>
```

PROC TEMPLATE: Not Just for Geeks Anymore

As noted previously, the item store contains templates for tables and styles. It also contains the tagsets used for the ODS MARKUP statement. SAS supplies a simple XHTML template, or you can create your own by running the following program:

Example 3.19 XHTML Template

```
proc template;

define tagset Tagsets.xhtml / store = SASUSER.TEMPLAT;
  define event cell_is_empty;
  put %nrstr(" ");
end;

define event doc;
  start:
      put "<?xml version=""1.0"" encoding=""UTF-8""?>" NL;
      put "<!DOCTYPE html PUBLIC ""-//W3C//DTD XHTML 1.1//EN""" NL;
      put """http://www.w3.org/TR/xhtml1/DTD/xhtml11.dtd"">" NL;
      put "<html xmlns=""http://www.w3.org/1999/xhtml"">" NL;
      ndent;
  finish:
      xdent;
      put "</html>" NL;
end;

define event doc_head;
  start:
      put "<head>" NL;
      ndent;
  finish:
      xdent;
      put "</head>" NL;
end;

define event doc_body;
  start:
      put "<body>" NL;
      put TITLE;
  finish:
      /* add W3C logo to page */
      put '<p>' NL;
      put '<a href="http://validator.w3.org/check/referer">' NL;
      put '<img src="http://www.w3.org/Icons/valid-xhtml11"' NL;
      put 'alt="Valid XHTML 1.1!" height="31" width="88" /></a>' NL;
      put '</p>' NL;
      put '</body>' NL;
end;

define event doc_title;
```

```
      put "<title>";
      put "SAS Output" / if !exists(VALUE);
      put VALUE;
      put "</title>" NL;
end;

define event proc_title;
   put "<h2>" VALUE "</h2>" CR;
end;

define event system_title;
   put "<h1>" VALUE "</h1>" CR;
end;

define event system_footer;
   put "<h1>" VALUE "</h1>" CR;
end;

define event byline;
   put "<h2>" VALUE "</h2>" CR;
end;

define event note;
   put "<h3>" VALUE "</h3>" CR;
end;

define event fatal;
   put "<h3>" VALUE "</h3>" CR;
end;

define event error;
   put "<h3>" VALUE "</h3>" CR;
end;

define event warning;
   put "<h3>" VALUE "</h3>" CR;
end;

define event table;
   start:
       put "<table border=""1"">" NL;
       ndent;
   finish:
       xdent;
       put "</table>" NL;
end;

define event row;
   start:
       put "<tr>" NL;
       ndent;
   finish:
       xdent;
       put "</tr>" NL;
end;

define event table_head;
   start:
```

```
        put "<thead>" NL;
        ndent;
    finish:
        xdent;
        put "</thead>" NL;
end;

define event table_body;
    start:
        put "<tbody>" NL;
        ndent;
    finish:
        xdent;
        put "</tbody>" NL;
end;

define event table_foot;
    start:
        put "<tfoot>" NL;
        ndent;
    finish:
        xdent;
        put "</tfoot>" NL;
end;

define event rowcol;
    putq " rowspan=" ROWSPAN;
    putq " colspan=" COLSPAN;
end;

define event header;
    start:
        put "<th";
        trigger rowcol;
        put ">";
        put VALUE;
    finish:
        put "</th>";
end;

define event data;
    start:
        put "<th" / if cmp( section , "head" );
        put "<td" / if !cmp( section , "head" );
        trigger rowcol;
        put ">";
        put     VALUE;
    finish:
        put "</th>" NL / if cmp( section , "head" );
        put "</td>" NL / if !cmp( section , "head" );
end;
```

```
mapsub = %nrstr("/&lt;/&gt;/&amp;/&quot;/");
map = %nrstr("<>&""");
split = "<br />";
output_type = "xml";
indent = 3;

end; /* define tagset */
```

PROC TEMPLATE listings can be intimidating, and many otherwise brave SAS users have avoided this procedure because of its reputation for complexity. No worries mate! It is just a question of putting together simple elements, as the preceding example illustrates.

There is a DEFINE statement for each event in the ODS output (shown in **bold** in the example) that describes what to do in that case. Consider the following extract:

```
define event doc_head;
start:
  put "<head>" NL;
  ndent;
finish:
  xdent;
  put "</head>" NL;
end;
```

All this says is that when ODS gets to the top of the document, it should write out a header. At the start of the header, put the opening tag <head>. After writing the initial tag, insert a line break and then indent. At the finish, stop indenting, put the closing tag </head>, and then put in a line break. See, that's not so hard!

In any case, pasting this code into the SAS editor window and submitting it should result in the following log message:

NOTE: TAGSET 'Tagsets.Xhtml' has been saved to: SASUSER.TEMPLAT

You should also see a new tagset called XHTML in SASUSER.TEMPLAT in the Templates window (see Display 3.10). You can now run Example 3.18 to get the result shown previously.

In conclusion, XHTML is an alternative to HTML for data display when used with custom stylesheets to create a formatted document in the client's browser. Using ODS, it is possible to customize your output for almost any conceivable application. Thus far, however, we have only considered static HTML output. In the chapters that follow, the much more interesting (and complex) topic of dynamic Web pages will be considered.

References

SAS Training
URL references are current as of the date of publication.

- SAS Institute Inc. *Advanced Output Delivery System Topics* (Course Notes). Cary, NC: SAS Institute Inc. http://support.sas.com/training/us/crs/odsadv9.html.

- SAS Institute Inc. *The Complete Guide to the SAS Output Delivery System.* Cary, NC: SAS Institute Inc. http://support.sas.com/rnd/base/early-access/odsdoc2/sashtml/tw5195/index.htm.

- SAS Institute Inc. *Creating Markup Language Files Using ODS Markup, SAS XML Libname Engine, and the TEMPLATE Procedure* (Course Notes). Cary, NC: SAS Institute Inc. http://support.sas.com/training/us/crs/lwxml.html.

- SAS Institute Inc. *Getting Started with the Output Delivery System Topics* (Course Notes). Cary, NC: SAS Institute Inc. http://support.sas.com/training/us/crs/odsgs.html.

- SAS Institute Inc. *Using the Output Delivery System to Create XML Files* (Course Notes). Cary, NC: SAS Institute Inc. http://support.sas.com/training/us/crs/odsxml.html.

SAS Publications

- Barlow, Todd. 2002. "Designing Web Applications: Lessons from SAS User Interface Analysts." *Proceedings of the Twenty-Seventh Annual SAS Users Group International Conference.* Cary, NC: SAS Institute Inc.

- Bessler, LeRoy. 2002. "Inform and Influence with Image and Data: Communication-effective Web Design for ODS, SAS, and SAS/GRAPH." *Proceedings of the Twenty-Seventh Annual SAS Users Group International Conference.* Cary, NC: SAS Institute Inc.

- Bessler, LeRoy. 2003. "Web Communication Effectiveness: Design and Methods to Get the Best Out of ODS, SAS, and SAS/GRAPH." *Proceedings of the Twenty-Eight Annual SAS Users Group International Conference.* Cary, NC: SAS Institute Inc.

- Bessler, LeRoy and Francesca Pierri. "Show Your Graphs and Tables at Their Best on the Web with ODS." *Proceedings of the Twenty-Seventh Annual SAS Users Group International Conference.* Cary, NC: SAS Institute Inc.

- Carpenter, Art. 2002. *Carpenter's Complete Guide to the SAS Macro Language.* 2nd ed. Cary, NC: SAS Institute Inc.

- Gilbert, Jeffery D. 2005. "Web Reporting Using the ODS." *Proceedings of the Thirtieth Annual SAS Users Group International Conference.* Cary, NC: SAS Institute Inc.

- Gupta, Sunil K. 2003. *Quick Results with the Output Delivery System (Art Carpenter's SAS Software Series).* Cary, NC: SAS Institute Inc.

- Gupta, Sunil K. 2001. "Using Styles and Templates to Customize SAS ODS Output." *Proceedings of the Twenty-Sixth Annual SAS Users Group International Conference.* Cary, NC: SAS Institute Inc.

- Haworth, Lauren E. 2001. *Output Delivery System: The Basics.* Cary, NC: SAS Institute Inc.

- Haworth, Lauren E. 1999. *PROC TABULATE By Example.* Cary, NC: SAS Institute Inc.

- Johnson, Bernadette. 2003. *Instant ODS: Style Templates for the SAS Output Delivery System.* Cary, NC: SAS Institute Inc.

- Lafler, Kirk Paul. 2002. "Output Delivery Goes Web." *Proceedings of the Twenty-Seventh Annual SAS Users Group International Conference.* Cary, NC: SAS Institute Inc.

- Morgan, Derek and Steve Hoffner. 2004. "Generating a Detailed Table of Contents for Web-Served Output." *Proceedings of the Twenty-Ninth Annual SAS Users Group International Conference.* Cary, NC: SAS Institute Inc.

- Newkirk, Gregory. 2004. "Using ODS, an Easy Approach in Creating HTML Web Pages." *Proceedings of the Twenty-Ninth Annual SAS Users Group International Conference.* Cary, NC: SAS Institute Inc.

- Olinger, Chris. 2000. "ODS for Dummies." *Proceedings of the Twenty-Fifth Annual SAS Users Group International Conference*. Cary, NC: SAS Institute Inc.

- Olinger, Chris. 2005. "A Sample Web-Based Reporting Container for ODS XML Reports." *Proceedings of the Thirtieth Annual SAS Users Group International Conference*. Cary, NC: SAS Institute Inc.

- Olinger, Chris. 1999."Twisty Little Passages, All Alike – ODS Templates Exposed." *Proceedings of the Twenty-Fourth Annual SAS Users Group International Conference*. Cary, NC: SAS Institute Inc.

- Pratter, Frederick. 2001. "Beyond HTML: Using the SAS System Version 8.2 With XML." *Proceedings of the Twenty-Sixth Annual SAS Users Group International Conference*. Cary, NC: SAS Institute Inc.

- Rafiee, Dana and Stan Andermann. 2001. "Simple Ways to Publish Reports and Graphs on the Web." *Proceedings of the Twenty-Sixth Annual SAS Users Group International Conference*. Cary, NC: SAS Institute Inc.

- Smith, Curtis A. 2002. "Web Enabling Your Graphs With HTML, ActiveX, and Java Using SAS/GRAPH and the Output Delivery System." *Proceedings of the Twenty-Seventh Annual SAS Users Group International Conference*. Cary, NC: SAS Institute Inc.

Links

- Creating Customized Tagsets to Use with ODS and XML LIBNAME Engine – http://support.sas.com/rnd/base/topics/odsmarkup/tagsets.html

- Extensible Markup Language – http://www.w3.org/TR/REC-xml.html

- HTML Data Set Formatter Syntax Reference – http://support.sas.com/rnd/web/intrnet/format/ds/dssyn.html

- HTML Formatting Tools – http://support.sas.com/rnd/web/intrnet/format/

- HTML Output Formatter Syntax Reference – http://support.sas.com/rnd/web/intrnet/format/out/outsyn.html

- HTML Tabulate Formatter Syntax Reference – http://support.sas.com/rnd/web/intrnet/format/tab/tabsyn.html

- Output Delivery System – http://support.sas.com/rnd/base/index-ods-resources.html

- PROC TEMPLATE FAQ – http://support.sas.com/rnd/base/topics/templateFAQ/Template.html

- Using ODS to Export Output in a Markup Language – http://support.sas.com/rnd/base/topics/odsmarkup/

- XHTML – http://www.w3.org/TR/2001/REC-xhtml11-20010531

Part 2

Access to SAS with SAS/IntrNet Software

Chapter 4

Remote Access to SAS

Client/Server Computing

In the 1970s and 1980s, data processing (or *information technology*, as it had become known) began to move away from a mainframe-based model to a network-centric model. Minicomputers such as the DEC PDP-8 and, later, scientific workstations and PCs made it possible to shift much of the computing power from one big central computer to distributed systems. Unfortunately, for projects where more than one person was working on the same task, the data still had to be shared. This led to two different scenarios, both fraught with potential problems.

One approach was for each user to get a separate copy of the data on his or her own local machine. Unfortunately, this approach increased the likelihood of the copies getting out of sync. The other way to manage sharing the data was to store the data on a central file server. When a user needed access to shared data, the entire data set had to be copied over the network to that user's system. For example, if you wanted to analyze only redheaded, left-handed, male violinists, the file server would send the whole file to your computer, where a selection program would then throw away all the observations that did not meet the criteria.

Obviously, this led to a lot of unnecessary network traffic, but in the early days of SAS it was the only way to manage shared data resources. The solution that was adopted is called the *client/server* model, because the distributed local workstations act as a *client* to a remote central host or *server*.

In this model, a program runs continually on the file server; the task of this program is to listen for client requests. When a data request comes in, the server program accesses the database, selects the desired records (all redheaded, left-handed, male fiddlers), and sends only those records across the network to the client for subsequent analysis. Alternatively, the server may also do some of the processing and send only summary tables to the client. In either case, the goal is to minimize the amount of network traffic by offloading computing responsibilities to the server.

This approach is clearly much more efficient and rapidly became the dominant paradigm for network computing. SAS software is unusual in that it offers several different data access methods, summarized in Table 4.1:

Table 4.1 SAS Data Access Methods

Access Method	Data	Processing
PC file server	Local client or network server	Local client
SAS/SHARE	Remote SHARE server	Local client
SAS/CONNECT	Remote CONNECT server	Remote CONNECT server
SAS Integration Technologies	IOM server	Distributed clients
SAS OPEN OLAP server	OLAP cubes	Distributed access to OLAP data

In addition to the original client/server components SAS/SHARE and SAS/CONNECT, SAS 8 introduced the *Integrated Object Model* (IOM) protocol:

> The Integrated Object Model (IOM) in SAS Integration Technologies provides distributed object interfaces to base SAS software features such as the procedural scripting language, data, file system, results content, and formatting services. IOM enables you to use industry-standard languages, programming tools, and communication protocols to develop client programs that access these services on IOM servers.
> (http://support.sas.com/rnd/itech/doc/dist-obj/iom.html)

Since the OLAP server is licensed as a separate product, in this chapter we shall focus on only the connection methods included with SAS AppDev Studio, which provide basic support for SAS remote data access. See http://www.sas.com/technologies/dw/storage/mddb/ for more information on this product as well as the SAS Web OLAP Viewer for Java and the Web OLAP Viewer for .NET.

This chapter presents examples of how to connect to a remote SAS server using SAS/SHARE and SAS/CONNECT. Chapter 10, "Using the SAS Open Metadata Architecture with the Integrated Object Model," describes the requirements for connecting to an IOM server.

Remote Data Services with SAS/SHARE

SAS/SHARE is used to manage multiple, simultaneous connections to a remote SAS server. In order to implement the client/server model using SAS/SHARE, several preliminary steps are required:

1. A TCP service must be configured on both the server and the local client.
2. The SAS/SHARE server must be configured to allow remote TCP/IP connections.
3. The server must be started.

Editing the TCP Services File

TCP services are defined in the **services** file, the location of which is operating system specific:

- Windows 2000 – `C:\winnt\system32\drivers\etc\services`
- Windows XP – `C:\windows\system32\drivers\etc\services`
- UNIX/Linux – `/etc/services`

Note that to establish a TCP connection, the service name must be defined on both the client and the server. Consequently, in order to connect to a UNIX or Windows host server from a Windows client, as shown in the examples that follow, it is essential that the service name and port number are added to the service files on each end.

The format of the **services** file is the same on all three platforms; the following example shows a portion of a Windows **services** file:

Display 4.1 Windows Services File

```
# Copyright (c) 1993-1999 Microsoft Corp.
#
# This file contains port numbers for well-known services defined by IANA
#
# Format:
#
# <service name>  <port number>/<protocol>  [aliases...]  [#<comment>]
#

echo            7/tcp
echo            7/udp
discard         9/tcp       sink null
discard         9/udp       sink null
systat          11/tcp      users               #Active users
systat          11/tcp      users               #Active users
daytime         13/tcp
daytime         13/udp
```

```
                    [... many lines omitted ...]

#SAS/CONNECT service
spawner                4016/tcp      # UNIX or OS/390 spawner

shr1                   5010/tcp      # local SAS server
shr2                   5011/tcp      # SAS/SHARE SERVER
```

Note that in general two kinds of services are available: tcp and udp.[1]

As the comment at the beginning of the code indicates, the function of the **services** file is to map port numbers to services. The last line of the example

```
shr2    5011/tcp      # SAS/SHARE SERVER
```

configures a service called shr2 on port number 5011. (The text following the # sign is just a comment labeling the service.) This example shows three different TCP connections for SAS services, one for a SAS/CONNECT spawner, one for local connections, and a third for managing SAS/SHARE access. As we shall see shortly, at least one TCP connection is required for using a remote SAS server.

Configuring TCP Security

The second requirement for a remote SAS/SHAR*E* session is that secure TCP access should be allowed by the SAS configuration file **sasv9.cfg**. Otherwise, there is no way to protect the SAS remote host against unauthorized access. By default, the configuration file is available in the SAS installation root directory, but editing this file should always be approached with trepidation. If you mess it up, you won't be able to start SAS, so make a backup before attempting to modify the configuration file.

In order to allow secure TCP/IP connections, you need to add the following lines to the configuration file:

```
-set TCPSEC _SECURE_
-set AUTHENCR REQUIRED
```

This will require all users to supply a valid password on the remote system before SAS will allow access to the data set. (Note: UNIX systems are case sensitive. Be sure to use upper case for these options on UNIX or Linux; Windows doesn't care.)

The option COMAMID=TCP must be specified, either at the system level or in the program itself. This allows SAS to connect to the server via the TCP/IP protocol. (COMAMID stands for *Communication Access Method ID*.) It is possible to set this option globally from the SAS Display Manager **System Options** menu, as shown in Display 4.2:

---

[1] UDP, the Universal Datagram Protocol, provides very little in the way of error correction and is used primarily for broadcast messages, such as DNS requests. SAS uses the more reliable TCP connection.

Display 4.2 Setting the COMAMID System Option Globally

Starting and Stopping the SAS/SHARE Server

Finally, the server must be started up on the remote system. Remember that a database engine is a
background process that runs continually on the server listening for connection attempts.
Consequently, a SAS data server program has to be running in order for the clients to have
something to connect to. PROC SERVER is used to start the server, as illustrated in the following
example:

Example 4.1 Sample Program to Start the SAS/SHARE Server

```
*******************************************************;
*****    start SAS/SHARE server on local host   ******;
*****    system administrator password: system  ******;
*****    user password: user                     *****;
*****    authentication is on                    ******;
*******************************************************;

proc server
  id=shr1
  oapw=system
  uapw=user
  authenticate=required;
run;
```

Four parameters are illustrated:

- `id` – the required service name defined in the `services` file.
- `oapw` – an optional administrator password; if included it must be supplied to stop the server.
- `uapw` – an optional user password; if included it must be supplied to connect to the server.
- `authenticate` – a choice of either `optional` or `required`; if `required` is specified, the user must supply a valid user ID and password in order to connect to the server, and TCPSEC must be set in the configuration file as illustrated in the previous section.

To stop the server, the OPERATE procedure is used:

Example 4.2 Stop the SAS/SHARE Server

```
*****************************************************;
*****   stop SAS/SHARE server on local host   *****;
*****************************************************;
proc operate serverid=shr1 sapw=system uid=_prompt_;
     stop server;
run;
```

Three parameters are included on the procedure statement:

- `serverid` – the required service name defined in the `services` file.
- `sapw` – an operator password specified when the service was started (the same as the `oapw` in the startup script).
- `uid` – a valid login name and password for the system that must be supplied if authentication is required. The argument value `_prompt_` displays a pop-up window for user ID and password; otherwise the value must be hardcoded in the program file, which is a security risk.

PROC OPERATE takes a number of options including `stop` (shown in Example 4.2), `display server` (show the status of a named service), `quiesce` (refuse any new connections), and start (restart a quiesced service).

The process for starting a server is different on UNIX and Windows, although the underlying concept is the same.

Assuming that the program shown in Example 4.2 has been saved as `sharesrv.sas`, you would start the server on UNIX by typing the following shell command:

```
nohup sas -noterminal -rsasuser -sysin sharesrv.sas &
```

Running the command with `nohup` (no hang up) ensures that you can log out and the server will continue to run. Appending an ampersand (`&`) causes this program to run in the background. (It is of course possible to add the script to the **init.d** directory so that it will be automatically started when the server reboots. You will need to convince your system administrator to do this for you, unless of course you are the sysadmin.)

Managing SAS/SHARE as a Windows Service

SAS recommends that the server be installed as a Windows service so that it will automatically restart if the system needs to be restarted. There are three ways to accomplish this, depending on the combination of SAS components available on the server.

SAS Service Configuration Utility

Note that this utility is available only on Windows NT/2000/XP; it is not available for UNIX or Windows 95/ 98/ME, since these do not support the concept of Windows application services. Once this tool has been installed, go to **Start ▶ Programs ▶ SAS ▶ SAS 9.1 Utilities ▶ SAS Service Configuration Utility**. Something like the following screen should appear:

Display 4.3 Installing a SAS/SHARE Service Using the SAS Service Configuration Utility

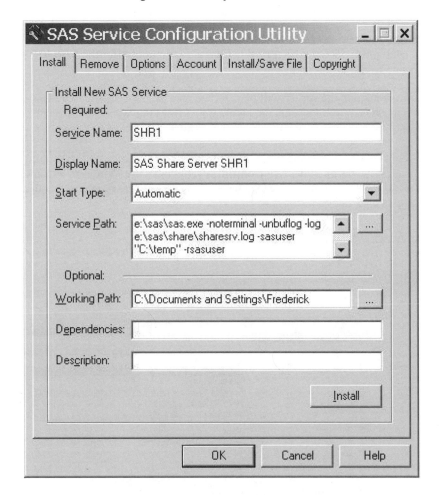

You need to fill in only the first four parameters (see the documentation available by clicking the **Help** button on the form):

1. **Service Name** - the internally registered name of the service being installed.
2. **Display Name** - the name of the installed service as it is displayed.

3. **Start Type** - the method for starting the service; the options are **Manual** or **Automatic**. Selecting the latter will start the service when the Windows server is rebooted; otherwise the service must be started from the Service Control Manager (see Display 4.4 below).

4. **Service Path** – the path to the location of the SAS program to start the service, plus any command-line parameters required by the service.

The only hard one is the **Service Path**. In Display 4.3, the SAS executable is located in the directory **E:\sas** and the server startup program is in **E:\sas\share**. Consequently the arguments to the service path are the following:

```
E:\sas\sas.exe -noterminal -rsasuser -sasuser "C:\temp"

-unbuflog -log E:\sas\share\sharesrv.log

-sysin E:\sas\share\sharesrv.sas
```

In other words, run the SAS program **sharesrv.sas** to start the server, do not display a terminal window, open a temporary SASUSER.PROFILE catalog, do not buffer the SAS log, and write the log out to the same folder as the input.

Click **OK** in the screen and the server will be installed as a Windows service. In this way it will be started when the system reboots and can be managed from the Windows Service Control Manager console.

If the service does not start automatically, it can be started manually from the Windows Services Manager application. Right-click the **My Computer** icon and select **Manage ▶ Services and Applications ▶ Services** and look for **SAS Share Server SHR1** as shown in Display 4.4. Click **Start the service** to start the Share server.

Display 4.4 Windows Service Control Manager

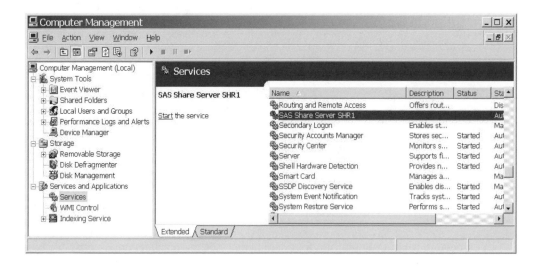

SAS AppDev Studio Service Manager

If SAS AppDev Studio has been installed on the server machine, another way to start the server on Windows is from **Start ▶ Program ▶ SAS AppDev Studio ▶ Services ▶ SAS AppDev Studio Service Manager**. Just check the desired boxes and click **Launch Services** to start the configured servers as shown in Display 4.5.

Display 4.5 SAS AppDev Studio Service Manager

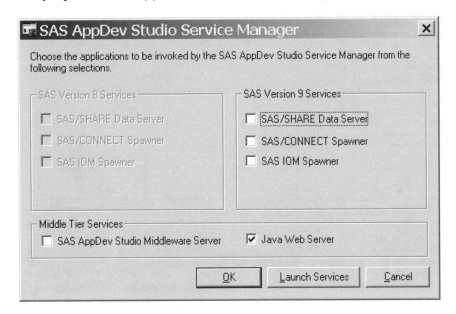

Managing Servers with SAS Management Console

The Open Metadata Architecture introduced in SAS®9 provides an alternative way to manage servers; see Chapter 10, "Using the SAS Open Metadata Architecture with the Integrated Object Model," for more information about using the SAS Management Console Server Manager. For now, it is sufficient to note that the Release 9.1 Configuration Wizard can be used to create several remote data servers, including SAS/SHARE and SAS/CONNECT servers.

Using SAS Management Console is probably the best way to manage remote servers—SAS recommends that your site be organized in this way—but there are so many options and different paths through this software that this process is best left to those responsible for managing organizational resources. An investment in learning this application is well worth the effort, but it is not necessary for end users. Display 4.6 illustrates an example of the default configuration; see "Introduction to SAS Management Console" at http://support.sas.com/publishing/pubcat/chaps/59903.pdf for an introduction to this product.

Display 4.6 SAS Management Console Server Manager

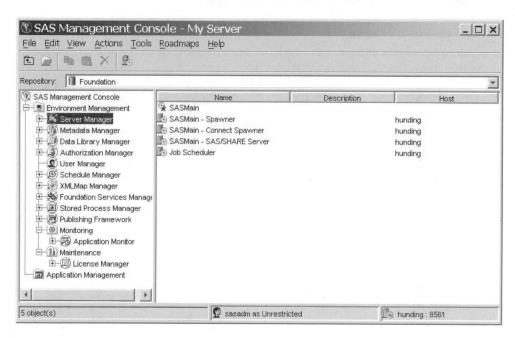

Access to Remote Library Services

Once the server is started, running a remote access program is simply a matter of specifying the correct LIBNAME options. Example 4.4 utilizes SAS/SHARE to implement a true client/server approach.

Example 4.4 Remote Library Services

```
libname SHARED
  slibref=SASHELP
  server=hygelac.shr1
  sapw=user
  passwd=_prompt_;

proc print data=SHARED.RETAIL;
  title "Retail Sales Total by Month: 1991-1994";
  where YEAR gt 1990;
  var MONTH SALES;
  id YEAR;
run;
```

The SAS LIBNAME SHARED references the SAS directory on the remote Linux server where the RETAIL data set is stored.

Four keyword parameters are included on this statement:

- `slibref` – a previously allocated remote SAS LIBNAME that will be mapped to the local LIBNAME `SHARED`; alternately, you can also specify a remote directory such as **`/usr/local/SAS_9.1/sashelp`**.

- `server` – a two-level name for the remote service; the first level is the *server name* (specified in the **hosts** file) followed by the *service name* defined in the **services** file.

- `sapw` – the user password specified when the service was started (`uapw`).

- `passwd` – the password for the specified login; using `_prompt_` will avoid hardcoding the password in the program.

The SAS client program sends a request to the server on the remote system. The server selects the observations requested by the statement `where YEAR gt 1990` and sends them to the client, where the PRINT procedure executes. (The output is not shown, since it is simply the result of the PRINT statement.) The client/server approach eliminates the need to send any unnecessary observations over the network connection between the two systems. In addition, SAS/SHARE allows *concurrent updates*; more than one user can write to the shared data file at the same time.

If SAS Management Console has been used to configure the server, the LIBNAME statement shown previously can be modified to use the new feature in SAS®9 that allows you to specify a port number rather than a named service. Note in this case that it is not necessary to add this port to your services file; the Server Manager has taken care of that for you. The following example replaces the LIBNAME statement shown in Example 4.4, and connects to a remote data library on a Windows server. The port number 8551 is prefixed by two underscores to indicate the specific TCP/IP connection.

```
libname SHARED slibref=SASHELP server=hunding.__8551;
```

The resulting output is the same as for Example 4.4.

Remote SQL Pass-Through (RSPT)

In addition to specifying the server on the LIBNAME statement, as in Example 4.4, it is also possible to use PROC SQL to connect to a remote server, as the following example illustrates:

Example 4.5 Remote SQL Pass-Through

```
proc sql;

  connect to remote
      (server=hygelac.shr1 sapw=user
       user=frederick password=_prompt_);

  select * from connection to remote
      (select YEAR, MONTH, sum(SALES)
            format=dollar12. label='Total Sales'
       from SASHELP.RETAIL
       group by YEAR, MONTH);
quit;
```

The first SQL statement establishes a connection to the remote server. The parameters on the CONNECT TO REMOTE statement are the same as we have seen previously in Example 4.4.

The SELECT statement downloads the entire result set from the parenthesized query. This second query is referred to as SQL *pass-through*, since the SELECT statement is executed on the server and only the results are returned to the client. Compare this statement to the following alternative:

```
select YEAR, MONTH, sum(SALES)
   format=dollar12. label='Total Sales'
   from connection to remote
       (select * from SASHELP.RETAIL)
   group by YEAR, MONTH;
```

In this version of the program, all of the data is transferred from the server, and the summation is done on the client. The advantage is that this reduces the load on the server; the drawback is of course that a lot of unnecessary records are copied over the network connection.

Remote Compute Services with SAS/CONNECT

The SAS/CONNECT product relies on an entirely different approach, essentially ignoring the processing resources of the local computer. Instead, the client serves only as a kind of post office, sending the program over the network where it runs entirely on the remote machine.

As with SAS/SHARE, using SAS/CONNECT requires that a server program be started on the remote system. This program is called a *Remote Host Spawner*. The spawner program listens for a connection to the host, just like a Web server. (Do not confuse the Remote Host Spanner with the Object Spawner, which provides IOM services.)

Starting a Remote Host Spawner
Starting the spawner is handled differently on Windows and UNIX, although again the principle is the same across all platforms.

On Windows, the spawner program is stored in the SAS root directory. Just run

```
spawner.exe -comamid -install
```

from a command window to install the spawner as a service. For a complete list of the available options, see "Spawners and Files," in the SAS System Help contents under **SAS Products ▶ SAS/CONNECT ▶ Communications Access Methods**.

Note that SAS/CONNECT spawners can also be implemented using SAS Management Console. The default installation scripts will install the service if this product is licensed at your site; you do not need to run the **spawner.exe** program if you are using a metadata profile to manage your hosts.

On UNIX, the spawner program is installed by default under the SAS root directory as **$SASROOT/utilities/bin/sastcpd**. Just as with SAS/SHARE, the TCP port must be defined in the /etc/services file, as for example:

```
spawner          4016/tcp          # UNIX spawner port
```

Assuming that this service name has been defined, the remote host spawner can be started with the following command:

```
sastcpd -service spawner -shell &
```

The **sastcpd** program runs by default as a daemon, so it is not necessary to use nohup. As with the SAS/SHARE server, if you want it to restart when the server reboots, install the script under the init.d directory.

Signing On to a Remote SAS Server

SAS/CONNECT requires an initial SIGNON command to the remote server. This can be accomplished by submitting something like the following statements:

```
options remote=hygelac.spawner;
signon cscript="E:\sas\connect\saslink\tcpunix.scr";
```

The REMOTE= option is used to identify the specific network host and the service name. The SIGNON script files **tcpunix.scr** (for UNIX) and **tcpwin.scr** (for Windows) are supplied by the SAS/CONNECT installation. Both scripts are on the local client in the specified directory, along with the scripts for TSO, CMS, etc. Again, be sure the option COMAMID=TCP is specified so that SAS can use TCP/IP to communicate with the server.

You may need to modify the SIGNON script. For example, if you need to connect to a UNIX host and the SAS executable is not in your default PATH on that system, you will need to edit the **tcpunix.scr** script to add the absolute path to the startup options. Search for the string

```
type 'sas -dmr -comamid tcp -device grlink -noterminal ';
```

and modify the string as necessary (e.g., replace the sas command with /usr/local/bin/sas or wherever your startup may be). Be careful, since this string may occur more than once in the script; you should modify it anywhere that it appears.

You can also submit the SIGNON command from the SAS Display Manager; select the menu item **Run ▶ Signon**, and enter the script filename and remote session name as shown in Display 4.7.

Display 4.7 Signing On to a Remote SAS/CONNECT Server on a UNIX System

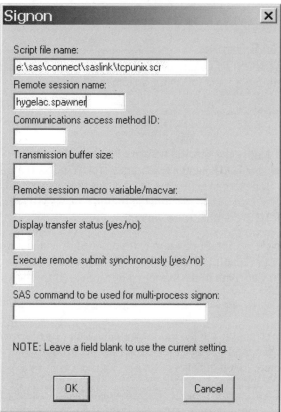

You can sign off from a SAS/CONNECT session by using **Run ▶ Signoff**. No arguments are necessary, since the default is to sign off from the currently active session.

Remote Compute Services

Following a successful signon, the RSUBMIT command is used to send the program to the remote system, where it executes as shown in the following example:

Example 4.6 Remote Compute Services

```
rsubmit;

proc tabulate data=SASHELP.RETAIL;
   title "Retail Sales In Millions Of $";
   class YEAR/descending;
   var SALES;
   table YEAR="" all="Total", SALES="" *
       (sum="Total Sales"*f=dollar8.
        pctsum="Overall Percent"*f=8.2
        n="Number of Sales"*f=8.
```

```
        mean="Average Sale"*f=dollar8.2
        min="Smallest Sale"*f=dollar8.
        max="Largest Sale"*f=dollar8.)/
       box="Year" rts=8;

   run;

   endrsubmit;
```

The results are generated remotely but displayed on the local system. The ENDRSUBMIT statement signals SAS to stop sending program statements to the remote system and start executing locally again. Note that on the procedure statement, the data set name SASHELP.RETAIL refers to the copy of the library on the remote server, not the local version.

In addition to Remote Compute Services, SAS/CONNECT also supports the following:

- **Remote Library Services** – similar to SAS/SHARE
- **Data Transfer Services** – upload and download of SAS data sets
- **Remote SQL Pass-Through** – similar to SAS/SHARE

The online documentation for SAS AppDev Studio includes a detailed explanation of when it is appropriate to use SAS/CONNECT as opposed to SAS/SHARE. Generally, SAS/CONNECT will do everything that SAS/SHARE can do and more, but the latter choice may be somewhat easier to deploy, since it has fewer bells and whistles.

Distributed Computing with the Integrated Object Model

The complexities of managing Integrated Object Model (IOM) servers are covered in detail in Chapter 10, "Using the Open Metadata Architecture with the Integrated Object Model." At this point, it is sufficient to note that IOM provides an alternative to SAS/SHARE and SAS/CONNECT for sites where these products are not available. The difference between IOM and the two older products is in the way the server can be accessed. There are three ways to connect to an IOM server:

1. using SAS Enterprise Guide to run SAS programs on the remote host
2. using SAS Stored Processes, either from SAS Enterprise Guide or from a Web application
3. directly using the Java, C++, or Visual Basic classes supplied by SAS

Since these components require some background in configuring and running the Open Metadata Architecture, discussion of the IOM server will be deferred until the appropriate chapters later in this book.

Like the dinosaurs, who grew feathers and evolved into birds, the client/server model has grown wings for the 21st century. In this new approach, called the *distributed object model*, objects are packaged as components such as JavaBeans and COM objects. These objects are then distributed across the network using technologies such as CORBA and DCOM. For developers who need to implement this level of functionality, *SAS Integration Technologies* are available as a comprehensive set of distributed object solutions; one would be hard pressed to find dinosaurs in North Carolina.

References

Client/Server

- Lowe, Doug, and David Helda. 1999. *Client/Server Computing for Dummies. 3rd ed.* Foster City, CA: IDG Books

- Orfali, Robert, Dan Harkey, and Jeri Edwards. 1995. *Client/Server Survival Guide. 3rd ed.* New York, NY: John Wiley & Sons.

SAS Documentation

URL references are current as of the date of publication.

- SAS Institute Inc. 2004. *Communications Access Methods for SAS/CONNECT 9.1 and SAS/SHARE 9.1.* Cary, NC: SAS Institute Inc. http://support.sas.com/apps/pubscat/bookdetails.jsp?catid=1&pc=58986

- SAS Institute Inc. 2004. *SAS/CONNECT 9.1 User's Guide.* Cary, NC: SAS Institute Inc. http://support.sas.com/apps/pubscat/bookdetails.jsp?catid=1&pc=58984

- SAS Institute Inc. 2004. *SAS/SHARE 9.1 User's Guide.* Cary, NC: SAS Institute Inc. http://support.sas.com/apps/pubscat/bookdetails.jsp?catid=1&pc=58985

- SAS Institute Inc. 2004. *SAS 9.1.2 Management Console: User's Guide.* Cary, NC: SAS Institute Inc. http://support.sas.com/apps/pubscat/bookdetails.jsp?catid=1&pc=59903

Chapter 5

Web Applications Programming

Server-Side or Client-Side?

All of the Web page examples shown in Chapter 3, "Creating Static HTML Output," are *static*. That is, each page is the output of one or more batch procedures, the results of which are fixed and unchanging. An ordinary HTML page does not allow *dynamic* Web content; some kind of additional scripting is required for the Web page to be interactive. In general, there are two ways to accomplish this:

- *client-side* Web programming using Java applets or JavaScript
- *server-side* Web programming using CGI, PHP, ASP.NET or Java servlets and JSP

The first approach is to create code that executes on the Web client. Programs of this type are stored on the server and are downloaded and run by the client, providing "just-in-time" applications services. The appendices to this book cover some of the strategies for providing SAS content on the client using DHTML (Appendix A, DHTML: Dynamic Programming on the Web Client with JavaScript) and Java Applets (Appendix B, Client-Side Programming with JavaApplets).

The main advantage of client-side programming is that it shifts the processing load from the Web server, allowing more simultaneous connections. However, there are several disadvantages as well:

- The Web application program has to be compatible with the client. Many older browsers do not support all JavaScript functions, and applets require the Java plug-in to function fully.

- Network traffic can increase as relatively large code segments are transferred to the client to be executed there.

Consequently, the decision as to where to locate Web applications programs depends on the type of application and the capabilities of the server and the network. Corporate Intranets may well have the bandwidth to support downloading applets, as well as the ability to make sure that all of the client systems have the necessary software installed. However, concern for these issues has led many developers to favor server-side applications programs for use on the World Wide Web.

A number of strategies are available for this, falling into two major groupings. The earliest approach to server-side Web programming was the *Common Gateway Interface* (CGI), included in the original NCSA HTTP Server. An alternative approach developed more recently is the use of embedded scripting languages. Of the latter, the most widespread are ASP.NET (Microsoft's Active Server Pages), PHP (Linux-based freeware), and JSP (JavaServer Pages). Part 4 of this book discusses the last of these in detail, and shows how webAF software can be used to create dynamic Web applications.

SAS/IntrNet software, included as part of SAS AppDev Studio 3.1, provides two Web development tools based on the older CGI technology:

- Application Dispatcher is a set of modules that allow parameters to be passed to SAS from Web pages.

- htmSQL provides access to SAS data from a Web page.

Chapter 6, "SAS Intr/Net: the Application Dispatcher," and Chapter 7, "SAS Intrnet: htmSQL," cover the use of each of these tools in turn. First, however, it is important to understand what CGI can and cannot do and how this affects Web development using SAS.

The Common Gateway Interface

Mosaic, the original HTML browser, was developed at the National Center for Supercomputing Applications (NCSA) located at the University of Illinois Champaign-Urbana. The NCSA scientists proposed a standard for serving dynamic Web content:

> The Common Gateway Interface (CGI) is a standard for interfacing external applications with information servers, such as HTTP or Web servers. A plain HTML document that the Web daemon **retrieves** is **static**, which means it exists in a constant state: a text file that doesn't change. A CGI program, on the other hand, is **executed** in real-time, so that it can output **dynamic** information.
> (http://hoohoo.ncsa.uiuc.edu/cgi/intro.html)

Note that CGI is not itself a programming language, any more than is HTML. CGI is simply a standard interface for programs that execute on the Web server. A Web page can include a request that asks the server to run a CGI script (written in C, Java, Visual Basic, or Perl) that supplies the requested service.

A CGI program is simply an applications program, installed on the server or accessible from it, that follows the rules required to work with the CGI standard. The most important of these rules specify where the program must be located, how it reads input parameters, and how it outputs the results.

One obvious problem with CGI scripts is one of security. As the NCSA documentation points out:

> Since a CGI program is executable, it is basically the equivalent of letting the world run a program on your system, which isn't the safest thing to do. Therefore, there are some security precautions that need to be implemented when it comes to using CGI programs. Probably the one that will affect the typical Web user the most is the fact that CGI programs need to reside in a special directory, so that the Web server knows to execute the program rather than just display it to the browser. This directory is usually under direct control of the webmaster, prohibiting the average user from creating CGI programs. There are other ways to allow access to CGI scripts, but it is up to your webmaster to set these up for you. At this point, you may want to contact them about the feasibility of allowing CGI access. (http://hoohoo.ncsa.uiuc.edu/cgi/intro.html)

As we shall see, a number of steps required to set up the SAS/IntrNet result from these security concerns. It is important to determine whether CGI access is allowed on your server at all; as pointed out previously, the Web administrator should be contacted to determine the appropriate directories, URLs, and permissions. In particular, running a CGI program on a server requires that the Web host be able to access the directory that contains the executable program. Usually, this directory is on the same physical platform as the Web server, although this is not required. The NCSA default was that the directory must be called **cgi-bin**, and all implementations of CGI since then have used this convention. (It is possible to customize the location of this directory so that more than one set of applications programs are available. How this is accomplished differs for Apache and IIS and is beyond the scope of this discussion; the developer is referred to the appropriate documentation.)[1]

SAS/IntrNet software is available in several versions for different operating system platforms. SAS has used system specific methods for each. On Linux and UNIX systems, Web applications programs are implemented as binary executables in the **cgi-bin** directory; on the Windows platform, in contrast, they are implemented as Windows **.exe** files. In addition, if Microsoft IIS is used as the Web server, SAS/IntrNet application programs are installed in the **scripts** directory, not **cgi-bin**.

---

[1] For the Apache HTTP server, see http://httpd.apache.org/docs-2.0/howto/cgi.html. Note that a number of security holes have been reported in installations using CGI scripts with IIS. Microsoft suggests converting from CGI to their proprietary ISAPI technology; see "Internet Programming Frequently Asked Questions" at http://msdn.microsoft.com/library/default.asp?url=/library/en-us/vccore98/HTML/_core_how_does_isapi_compare_with_cgi.3f.asp.

A CGI Example

In order to appreciate how SAS/IntrNet is implemented, you might find it useful to review some simple examples of server-side Web application programs. Most CGI programs are written in Perl, although this is not required. Perl (*Practical Extraction and Reporting Language*) has been around longer than the World Wide Web. It was originally developed as a UNIX scripting language, and it is still widely used for purposes that have nothing to do with Web programming, although it is sufficiently integral to Web development that it has been called "the duct tape of the Internet."
(http://www.unixreview.com/)

The ubiquity of Perl scripts is a result of a combination of three factors:

- Perl is available for free.
- Perl is cross-platform, running on a variety of operating systems and hardware.
- The standard Perl distribution has a built-in module, *CGI.pm*, to automate many common Web programming tasks.

The following examples are implemented as Perl scripts. A number of excellent online Perl tutorials are available. However, knowledge of Perl is not necessary for SAS/IntrNet developers; consequently, it will not be covered here.[2] Nevertheless, it is instructive to see how CGI programming is generally accomplished using Perl scripts to dynamically generate Web content on the server. Example 5.1 shows some of the features of CGI scripting in Perl:

Example 5.1 Hello World CGI Program

```
#!/usr/bin/perl
##
##  HelloWorld.pl - My first Perl program
##
use CGI ':standard';            # include CGI module
print header;                   # generate MIME content line
print start_html "CGI Examples"; # generate starting HTML tags

# print "Hello World"
print h1( { -style=>'color: blue; '}, 'Hello World!');

# generate date/time field
( $s, $m, $h, $d, $mm, $y ) = localtime(time);

$y+=1900;                       # convert 2-digit year to 4
$mm++;                          # months start with 0

print p( {-style=>'font-weight: bold; font-size: 24;'},
     "The time now: $h:$m on $mm/$d/$y:");

print end_html;                 # generate ending HTML tags
```

---

[2] See http://directory.google.com/Top/Computers/Programming/Languages/Perl/WWW/ Tutorials/, and see the references at the end of this chapter.

Perl syntax might seem familiar to SAS programmers. Semicolons are used as statement separators; most Perl programmers put them at the end of every statement. Coding is free format and the use of white space is encouraged. One important difference between Perl and SAS programming is that all Perl scalar variable names (including both numeric and character types) must begin with dollar signs ($). Comments start with pound signs (#) and can appear at the end of the line or as a separate statement. Perl PRINT statements are the equivalent of SAS PUT statements; they send any results to the standard output.

This program makes extensive use of Perl's built-in HTML generating tools (and none at all of Perl regular expressions, which have befuddled many a programmer). Six standard functions are illustrated:

- `header` – generates the MIME content-type line and is required if the output is to be displayed in a Web browser (generates line 1 in the previous example)
- `start_html` – generates the XHTML heading (lines 2-10)
- `end_html` – generates the XHTML footer (lines 14-15)
- `h1` – encloses the text in an HTML `<h1>` tag (line 11)
- `p` – text is enclosed in an HTML `<p>` tag (lines 12-13)
- `localtime(time)` – displays the current system time as an array of six numbers: seconds, minutes, hours, day of the month, month and year

Generally, CGI scripts are initiated by a client application (usually a Web browser) and execute on the Web server, which then directs the output back to the client. It is possible, however, to run this program from the command line, with the resulting output (modified slightly for readability):

Example 5.2 Perl-Generated HTML Source

```
Content-Type: text/html; charset=ISO-8859-1

<?xml version="1.0" encoding="iso-8859-1"?>
<!DOCTYPE html PUBLIC "-//W3C//DTD XHTML 1.0 Transitional//EN"
        "http://www.w3.org/TR/xhtml1/DTD/xhtml1-transitional.dtd">
<html xmlns="http://www.w3.org/1999/xhtml"
  lang="en-US" xml:lang="en-US">

<head>
  <title>CGI Examples</title>
</head>
<body>
  <h1 style="color: blue; ">Hello World!</h1>
  <p style="font-weight: bold; font-size: 24;">
  The time now is 10:44 on 12/22/2005.</p>
</body>
</html>
```

Specifying the Perl script filename in the URL causes the server to route the output of Example 5.1 to the client's browser, with the result shown below.

Display 5.1 Generated HTML Output

The URL http://<server-name>/cgi-bin/example5-1.pl references the default `cgi-bin` directory on the server. Perl is interpreted, not compiled, so all that is necessary is to copy the Perl program shown in Example 5.1 to this directory.

There are a number of system requirements in order for this program to work correctly:

1. The server must be configured to support CGI. On an Apache server, this means that **mod_cgi** must be enabled.
2. The default **cgi-bin** directory must be executable.
3. The Perl program file (Example 5.1) must be executable.
4. The first lines of the program (in this case **#!/usr/bin/perl**) must point to the correct path to the Perl executable.

Note that the original Perl source shown cannot be displayed in a Web browser. Selecting **View Source** in the browser window shows instead the generated HTML code illustrated in Example 5.2, with the exception of the first line, the MIME content-type, which is interpreted by the Web browser as an instruction to display the remaining lines as HTML.

Passing Parameter Values to Web Applications

Probably the most common use of CGI programs is to collect and process data supplied by the user and return the results as a Web page. For example, CGI scripts can be used to search the World Wide Web, save input to a database, display a graph, or return the output of a calculation based on input data.

In order to accomplish this, values collected on one HTML pages must somehow be passed to another. The solution is to add pairs of fields to the URL that references the new page. These Web application parameter elements are called *name/value pairs*. Each pair consists of a variable name, followed by the equals sign, followed by the variable value (just like an assignment statement). Multiple pairs can be separated by an ampersand (&). Name/value pairs are appended to the end of the URL following a question mark.

The following example illustrates one way that values can be passed. The Perl script shown in Example 5.3 creates a Web form to convert Centigrade temperatures to Fahrenheit and vice versa.

Example 5.3 Perl Temperature Conversion Calculator

```perl
#!/usr/bin/perl
##
## Sample Perl program for temperature conversion
##
use CGI ':standard';

print header;
print start_html "Perl Temperature Conversion Examples";

# get input parameters and calculate result
$temp = param("input");
$type = param("convert");

if ( $type == 1 )            # convert F to C
{
        $result = 5*($temp - 32)/9;
}
elsif ( $type == 2 )# convert C to F
{
        $result = 9*$temp/5 + 32;
}

print '<div style="text-align: center; ">';
print h1 ( { -style=>'color: blue; '},
   'Temperature Conversion Calculator' );
print '<form name="calculator" action="convert.pl" method="get">';
print p ( '<strong>Enter a temperature and select a conversion
type:</strong>',
   '<input type="text" name="input" value=', $temp,'>' );
print p ( '<input type="radio" name="convert" value="1">
   Fahrenheit to Centigrade' );
print p ( '<input type="radio" name="convert" value="2">
   Centigrade to Fahrenheit' );
print p ( '<input type="submit" value = "Submit"
            style="color: blue; font-weight: bold;">',
          '<input type="reset" value = "Clear"
            style="color: blue; font-weight: bold;">' );

if ($temp gt '') # "input" parameter not blank or missing
{
```

```
            print p ('<strong>Result:</strong>',
               '<input type="text" name="result" value=', $result, '>' );
      }

      print '</div></form>';
      print end_html;
```

The Perl program creates a dynamic HTML form called calculator. When the **Submit** button on the form is clicked, the values entered into the text box are assigned to the input parameter, and the value corresponding to the selected radio button is assigned to the convert parameter. Both values are appended to the URL as name/value pairs.

This program uses many of the same Perl HTML functions as Example 5.1, with one important addition: param(). The single argument to this function is the parameter name and the value returned is the corresponding value appended to the URL. In this example, the statement $temp = param("input") creates a new Perl variable called temp and assigns it the value of the input parameter, which is specified in the text box on the Web page. This may be made a little clearer by comparing the following two output screens:

Display 5.2 CGI Temperature Conversion Calculator (First Time)

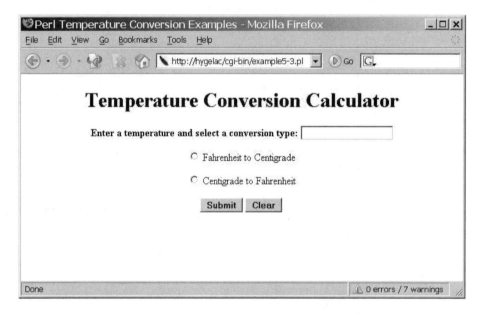

This is the initial state of the program. Now see what happens when the value "-40" is entered in the input text box, the **Fahrenheit to Centigrade** radio button is selected, and the **Submit** button is clicked. Since the value of the variable $temp is no longer null, the **Result** text box appears, with the computed answer: -40° F equals -40° C—information that will be of importance primarily to Minnesota users.

Display 5.3 CGI Temperature Conversion Calculator (Second Time)

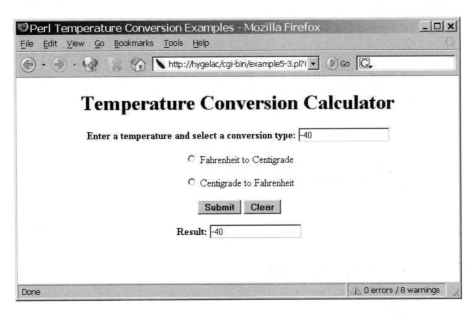

The URL now includes the appended parameters after a question mark character:

> http://hygelac/cgi-bin/example5-3.pl?input=-40&convert=1

These are the values from the form on the preceding Web page. For the current page and any subsequent ones, the URL will now include these name/value pairs.

One of the advantages of URL rewriting is that when a Web page is bookmarked, the name/values pairs are also stored. In this way a CGI program can be recalled at a later date without re-entering the form values. The main drawback to this approach is that it doesn't work very well for login screens, since the password would be appended to the URL and could be stored in the bookmark! There is an easy way around the problem, which is simply to change the form method from `get` to `post`. In the latter method, parameters are included in the page that is sent, but they are not displayed in the URL. If the page is bookmarked, however, the parameter values are not saved and the form information must be re-entered.

SAS makes use of name/value pairs to communicate with a SAS server. The methods available are quite powerful, but somewhat complex; they form the subject of the next two chapters on the Application Dispatcher and htmSQL.

References

Learning Perl

- Cassell, David L. 2001. "A Perl Primer for SAS Programmers." *Proceedings of the Twenty-Sixth Annual SAS Users Group International Conference.* Cary, NC: SAS Institute Inc.

- Lash, David A. 2002. *The Web Wizard's Guide to Perl and CGI.* Boston, MA: Addison-Wesley.

- Schwarz, Randal L., and Tom Phoenix. 2001. *Learning Perl. 3rd ed.* Sebastopol, CA: O'Reilly & Associates.

Links

URL references are current as of the date of publication.

- CGI on Apache – http://httpd.apache.org/docs-2.0/howto/cgi.html

- Common Gateway Interface – http://hoohoo.ncsa.uiuc.edu/cgi/intro.html

- Perl Tutorials – http://directory.google.com/Top/Computers/Programming/Languages/Perl/WWW/Tutorials/

- Using CGI on IIS – http://msdn.microsoft.com/library/default.asp?url=/library/en-us/vccore98/HTML/_core_how_does_isapi_compare_with_cgi.3f.asp

Chapter **6**

SAS/IntrNet: the Application Dispatcher

Overview

The first Web technology introduced by SAS was the SAS/IntrNet software product, a suite of CGI-based Web applications. The other SAS AppDev Studio components— webAF and webEIS— are used to develop Java applets and JavaServer Pages. In contrast, using SAS/IntrNet requires no Java or Perl programming and consequently is well-suited to sites where these programming resources may not be available.

SAS/IntrNet software provides two sets of CGI-based tools: Compute Services (Application Dispatcher) and Data Services (htmSQL). This chapter and the following one consider each of these in detail. In addition, several additional special-purpose components are available, including the SAS Design-Time Controls (see Appendix C,

"SAS/IntrNet: Design-Time Controls") and the SAS/CONNECT and SAS/SHARE drivers for JDBC.

The most complex of these both in terms of installation and use is the Application Dispatcher. Once the various software components have been properly configured, however, interacting with the Application Dispatcher requires no particular expertise on the part of the Web user. An investment in understanding the details of this product thus provides a relatively high level of functionality for a small ongoing resource outlay.

The Application Dispatcher itself consists of four components:

- *Input* – an HTML form to supply parameters
- *Program* – a SAS program to process the data from the form
- *Application Server* – a background SAS session to run the program
- *Application Broker* – a CGI program to pass the data from HTML to the Application Server

The latter two are supplied by SAS as part of SAS/IntrNet software; instructions for deploying them are included in the SAS online documentation.

The remaining two elements are user supplied. The input component is usually an HTML form, but it can be any element that includes a reference to the Application Dispatcher, such as a Java applet or ActiveX control. As the documentation notes:

> Regardless of the mechanism used, it must minimally send a service ID (passed in the _SERVICE field) and the name of the program (passed in the _PROGRAM field) to the broker (that is the CGI program).[1]

The rest of this chapter is an effort to illuminate this somewhat cryptic instruction.

Installing the Application Broker

In contrast to most of the SAS installation procedure, SAS/IntrNet installation requires some degree of foreknowledge.[2] Although most of the software is installed automatically by the SAS setup routines, the Application Broker software must be configured manually. The first puzzle is that the necessary software is included on the installation CD labeled "SAS Client-Side Components." This is somewhat peculiar, since the Application Broker is not a client application but a Web server one!

The SAS/IntrNet installation procedures have been revised somewhat for SAS®9, making the process more straightforward. First, open the Software Navigator from the "SAS Client-Side Components Volume 2" installation CD for SAS 9.1.3. For UNIX systems, this is accomplished by running the shell file **setup.sh**, while on Windows systems the corresponding file is **setup.exe**. Choosing SAS/IntrNet in the left-hand window opens the menu shown in Display 6.1.

---

[1] SAS Institute Inc., 2001. *Getting Started with SAS AppDev Studio* (Cary, NC: SAS Instistute Inc.), 31.
[2] The one-day class on SAS Web Tools: SAS/IntrNet Administration includes a session on configuring the Application Dispatcher. For a more general introduction, SAS also offers the course SAS Web Tools: Overview of SAS Web Technology.

Display 6.1 SAS Client-Side Components Installation Screen

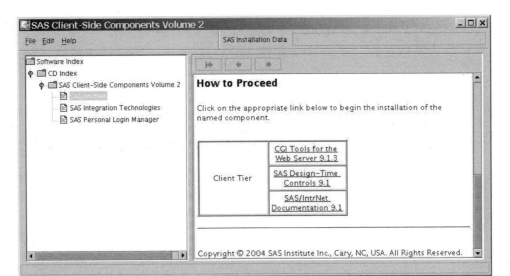

At a minimum, it is necessary to install the CGI Tools for the Web Server 9.1.3 package. (See Appendix C for a discussion of the Design-Time Controls.) Clicking on this link takes you to a list of choices that are specific to your operating system. Note that there is only one installation disk for all platforms. For Windows, the Application Broker and htmSQL are installed by running the **websrv.exe** program, which brings up an installation wizard. On a Linux or UNIX system, the Navigator application extracts a Web archive file to a specified installation directory; by default this is **/websrv_Linux/** under the SAS root directory.

In either case (Windows or UNIX) you will be asked three questions; you need to know the answers *before* you start the installation process. The dialog shown in Display 6.2 illustrates a typical Linux installation for the Apache Web server; since a Windows installation requires the same information, albeit in a different format, it is not shown. User inputs are shown in **bold**.

For the UNIX installation, first change to the resulting directory **/websrv_Linux/websrv** and then type

```
./INSTALL
```

Remember, UNIX and Linux are case sensitive, so be sure you type the name of the installation script in all uppercase. In addition, you should run the script as the `root` user or with `sudo` in order to have the needed file permissions.

Note that the installation documentation for Linux now includes an example of this dialog, along with an additional section on "Alternate Manual Installation." The automatic installation script seems to work fine, so you probably will not need to consider doing it manually.

Display 6.2 Installing the Application Broker for the Apache Web Server

```
Welcome to the SAS/IntrNet CGI Tools installation script.

Copyright © (2002-2003) SAS Institute Inc., Cary, NC, USA
All Rights Reserved

You will be prompted for information about your Web server
configuration and how you wish to install SAS/IntrNet. This
installation script will move the files that were extracted
from the CGI Tools package into a directory under your Web
server.  You will be able to review and confirm your responses
before any updates are made to your system.

The CGI Tools samples must be installed in the Web server
directory corresponding to the URL http://<your_server>/sasweb.

Enter the physical path corresponding to
http://<your_server>/sasweb.
If this directory does not exist, it will be created for you.

Path: /usr/local/apache2/htdocs/sasweb
/usr/local/apache2/htdocs/sasweb does not exist.

Do you wish to create it [Y]?  [Enter]
/usr/local/apache2/htdocs/sasweb created

Enter the physical path for SAS/IntrNet CGI executables. This
can be a standard CGI directory or a new directory reserved for
SAS/IntrNet.  Setup will create this directory for you if it
does not exist.

Path [/usr/local/apache2/cgi-bin]: [Enter]

The CGI executables and configuration files must be installed
in a Web server directory with execute privilege.  We suggest
using the URL http://<your_server>/cgi-bin for these files,
although you may use any URL.

Enter the URL path corresponding to the physical path
/usr/local/apache2/cgi-binCGI URL [http://<your_server>/
cgi-bin]:
http://hygelac/cgi-bin

The following steps will be performed:

- Sample files will be updated to use http://<your_server>/
cgi-bin for all CGI references.

- Sample files will be installed to
/usr/local/apache2/htdocs/sasweb/IntrNet9.
Existing files will be overwritten.

- CGI executables and configuration files will be installed to
/usr/local/apache2/cgi-bin.  Existing executables will be
renamed as a backup before the new executables are installed.
New configuration files are installed with a .cfg_v9 extension
so that existing configuration files are not overwritten.  If
```

```
no existing configuration file is found the .cfg_v9 file is
copied to create a new .cfg file.

- Java Graphics files will be moved to
usr/local/apache2/htdocs/sasweb/graph.
Existing files will be renamed as a backup.

Do you wish to continue [Y]? [Enter]

Modifying URL to CGI Tools in sample files
Moving samples to /usr/local/apache2/htdocs/sasweb/IntrNet9
Moving CGI files to /usr/local/apache2/cgi-bin
Moving Java Graphics files to
/usr/local/apache2/htdocs/sasweb/graph

The installation is complete.  Verify that your Web server is
configured to:

 - map http://<your_server>/cgi-bin to /usr/local/apache2/
   cgi-bin
 - allow CGI execution in http://<your_server>/cgi-bin
 - map http://<your_server>/sasweb to
   /usr/local/apache2/htdocs/sasweb

Once your SAS Servers are configured and started, you can view
SAS/IntrNet samples at
http://<your_server>/sasweb/IntrNet9/samples.html.
```

The first question the script asks is the path to the **sasweb** folder. Sample values for the two main types of Web server are as follows:

- Apache on UNIX: **$APACHE_HOME/htdocs/sasweb**, where **$APACHE_HOME** is the Apache server root directory

- Apache on Windows: **$APACHE_HOME\htdocs\sasweb**

- Microsoft Internet Information Services (IIS): **C:\Inetpub\wwwroot\sasweb**

Note that these directory paths are specified by the user at the time the Web server software is installed; the values illustrated are just some of the possible configurations.

The second question is the path to the **cgi-bin** directory. see Chapter 5, "Web Applications Programming," for a discussion of function and location of this directory. In the above example, the installation script correctly guessed it as **$APACHE_HOME/ cgi-bin**, so pressing the ENTER key accepts the default. On Windows the default path is **C:\Inetpub\scripts**.

The last question is the URL of the **cgi-bin** directory. If you chose all the defaults when performing the install, this is http://<server-name>/cgi-bin for an Apache Web server, while on a Windows system running IIS it is http://<server-name>/scripts instead.

Once all three path names are specified, the installation script copies the files from the CD to the specified directories as the preceding log above indicates. Unfortunately, we're not done yet; there are two more pieces to correctly configuring the Application Dispatcher software before we can run the sample programs supplied.

Creating an Application Dispatcher Service

Once the Application Broker has been installed, it is now necessary to configure and start at least one service. Logically enough, the default service is called "default." Since this is the service name used for the examples that are shipped with the software, it is a good idea to configure a service with this name. Depending on the requirements of the application, it may be useful to configure other services as well; see http://support.sas.com/rnd/web/intrnet/dispatch/defsvc.html for more information about how to create the default service.

Application services are created using an operating system-specific utility from SAS called *inetcfg*. The inetcfg program comes in four versions: Windows, UNIX/Linux, OpenVMS, and z/OS for IBM mainframes. Only the first two of these are covered here; for the other operating systems please refer to the SAS/IntrNet documentation.

Defining an Application Dispatcher Service for UNIX or Linux

The following log illustrates the process of creating the default service on a Linux system. The configuration script is written in Perl (see Chapter 5 for a description of the Perl programming language). You will need to modify the script by inserting a line at the top that points to the location of the Perl interpreter; see Example 5.1 for an illustration. On the `hygelac` system, this statement is the following:

```
#!/usr/bin/perl
```

Only one user-supplied value is required for installation: the name of the directory for serving applications, shown in **bold** in Display 6.3. Make a note of the pathname to this directory, because it's easy to lose track of, but is critical for administering the service.

The remainder of the prompts show the defaults selected by pressing the ENTER key after each. Note that the TCP/IP server port suggested is 5001. This port must be available; check the file **/etc/services** to make sure it is not in use for any other service. It does not need to be added to the services file.

Display 6.3 Creating the Default Application Service for Linux

```
Root directory for SAS/IntrNet services
(/home/frederick/IntrNet): /usr/local/apache2/intrnet
/usr/local/apache2/IntrNet does not exist
Do you wish to create it (Y): [Enter]

What kind of service do you wish to configure?

1 - Socket Service
2 - Pool Service
3 - Launch Service
4 - Load Manager
5 - Spawner

Enter service type? (1): [Enter]
```

Welcome to the SAS/IntrNet Application Dispatcher service
definition installation.

 Copyright © (2002-2003) SAS Institute Inc., Cary, NC, USA
 All Rights Reserved

This script will set up a service definition for the
SAS/IntrNet Application Dispatcher. You should know the type
of service you wish to create and the TCP port number(s) or
name(s) you wish to use (socket service only) before
continuing.

You will be asked to provide a directory name. The script will
create this directory and place server startup and log files in
it.

Name of the new service? (default): **[Enter]**

How many servers would you like for this service (maximum 20)?
(1): **[Enter]**

Please enter TCP/IP port values for this service. You may use a
service name such as appsrv or a port number such as 5001. If
you use a name, please remember to include this name in your
network services file along with a port number.

Port name or number for server 1 (5001): **[Enter]**

Do you want to protect the administration of this service with
a password? (N): **[Enter]**

```
Service name       : default
Service type       : Socket
Root Directory     : /usr/local/apache2/IntrNet/default
Number of servers  : 1
Server port(s)     : 5001
Admin password     : none
```

Create this service? (Y): **[Enter]**

The service directory has been created.

To start the default service execute
/usr/local/apache2/IntrNet/default/start.pl

The files necessary for this service have been created. To
complete the configuration perform the steps outlined in
this checklist.

* Install the Application Broker on your Web server machine.

* Create a service definition in your broker.cfg file for the
"default" service. For example:

```
SocketService default
  ServiceAdmin "[your-name]"
  ServiceAdminMail "[your-email]@[your-site]"
  Server      hygelac
  Port        5001
  FullDuplex  True
```

Defining an Application Dispatcher Service for Windows

The following example shows how to create the default Application Dispatcher service. Note that there are two different things going on here. First, an Application Dispatcher service is being created on the Windows platform. Then a Windows service is being defined that will automatically start the Application Dispatcher service that was just defined. For convenience, the following instructions are repeated from the documentation at http://support.sas.com/rnd/web/intrnet/dispatch/defsvc.html#windows:

Display 6.4 Creating the Default Application Service for Windows

```
Perform the following steps to create and start the default
service:

1. From the Start menu, select Programs ➤ SAS (or other program
   group where SAS is installed) ➤ IntrNet ➤ Create a New
   IntrNet Service.

   The IntrNet Config Utility Welcome window appears.

2. Read the information in the Welcome window, and then select
   Next to continue.

3. Select Create a Socket Service, then select Next to continue.

4. Type default as the name of the new service. Select Next to
   continue.

5. Specify the directory where you want the configuration utility
   to place your service directory and control files. The default
   location (under your SASUSER directory) is recommended. Select
   Next to continue.

6. Type the TCP/IP port number that you reserved for the default
   Application Dispatcher service. Select Next to continue.

7. A password is not necessary for the default service. You can
   add an administrator password later if you use this service
   for production applications. Select Next to continue.

8. The Create Service window displays all of the information that
   you specified for this service. Verify that the information is
   correct and then select Next to create the service.

9. Select Next and then Finish to complete the setup of the
   default service.
```

Note that specifying a name in step 4 other than "default" will create a service with that name. In that case, in step 6 don't use 5001 as the service number, because this is already in use for the default service; use another number, such as 5002.

Configuring the Default Application Service

Once the default service is defined, it must be configured manually. SAS/IntrNet services in SAS 8 and later are defined in the file **broker.cfg**, consisting of a set of directives in a specific format; the default location for the **broker.cfg** file on Windows systems is **C:\inetpub\scripts**. (In this regard it is like the **httpd.conf** file that is used to configure the Apache Web server software.) Editing the file **broker.cfg** is the same for UNIX and Windows; more information about broker configuration file directives is available at http://support.sas.com/rnd/web/intrnet/dispatch/bconfig.html.

The **broker.cfg** file is installed in the **cgi-bin** directory specified in Display 6.2; in the Linux example this is **/usr/local/apache2/cgi-bin**. Whether on UNIX or Windows, edit this file using your preferred text editor—*vi* or *emacs* for UNIX and Notepad or some third-party editor on Windows. (These instructions are also available at http://support.sas.com/rnd/web/intrnet/dispatch/addefsvc.html.) Search for the following section of the file:

Display 6.5 Application Broker Configuration for the Global Administrator

```
# Global administrator
# Used in error messages before a service has been established
or when
# the service administrator is omitted.
# May be overridden on a per-service basis with the
ServiceAdmin and
# ServiceAdminMail directives.
# ***> Modify this for your site <****
Administrator "Your Name"
AdministratorMail "yourname@yoursite"
```

Change "Your Name" and "yourname@yoursite" to the appropriate values for your site.

Next scroll down to find the following section:

Display 6.6 Application Broker Configuration File for the Socket Service

```
SocketService default "Default Socket Service"
   ServiceDescription "Default service used for
SAS/IntrNet samples."
   ServiceAdmin "Your Name"
   ServiceAdminMail "yourname@yoursite"
   Server appsrv.yourcomp.com
   Port 5001
#  Remove the following line for any servers before V8.1
   FullDuplex    True
```

You need to edit this file only to verify or change the values shown in bold: `ServiceAdmin`, `ServiceAdminMail`, `Server`, and the `Port` number if not 5001. The first two don't matter much, since they are just used to send you an e-mail if there's some kind of problem (and who wants that?). The third parameter, the server name, is crucial. Be sure to specify the correctly qualified DNS of your server, or else nothing else will work.

Starting and Stopping the Application Server

Finally, you need to start the Application Server. This is an operating system-dependent process.

On UNIX, change to the directory defined as the first step of the inetcfg process; in Display 6.2 this is **/usr/local/apache2/IntrNet**. There should be a directory for each service defined; at a minimum, one called **default**. Edit the file **start.pl** to begin with a reference to the correct path to the Perl interpreter, as discussed previously in regard to the **inetcfg.pl** script.

To start the service, you need to use an account that has root privileges. Change to the default directory and type

```
start.pl
```

It is not necessary to include `nohup` in this command; through the magic of forking processes, the script ensures that the service will continue to run even if you subsequently log out. If the server is rebooted, obviously it is necessary to restart the service; otherwise, it just runs as a background process with a parent PID of 1, indicating that it is a root daemon.

The start script initiates a SAS session on the server using the input file **appstart.sas**. We shall encounter this program again later; for the moment it is sufficient to note that this program uses PROC APPSRV to create the SAS batch environment that runs as a background process on the server.[3]

Starting the service is easier on Windows; go to the **Start** button and select **Programs ▶ SAS ▶ IntrNet ▶ default Service**. The folder **default Service** is the installation default directory for the default service. (If the last sentence makes sense to you, you're starting to catch on!) This folder contains shortcuts to several batch files in the Service Directory subfolder; these shortcuts are available from the Windows **Start** menu under **IntrNet ▶ default Service** as shown in Display 6.7.

---

[3] The APPSRV procedure is documented in the SAS Help and Documentation under "SAS/IntrNet Software" in "Help on SAS Software Products."

Display 6.7 SAS/IntrNet Installation Options for Windows

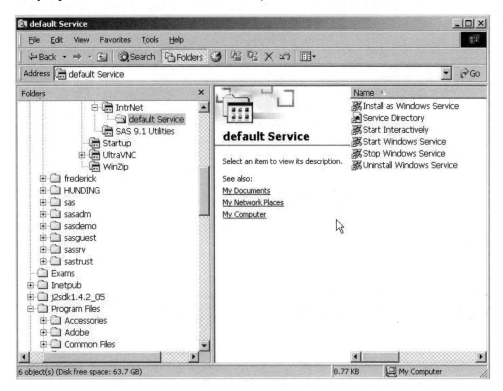

The shortcuts link to the following command files:

Table 6.1 SAS/IntrNet Shortcuts

Shortcut	Program	Description
Install as Windows Service	install.bat	Installs the default service as a windows service
Service Directory		Windows shortcut to the directory containing script files, logs and other files associated with the default service
Start Interactively	appstart.bat	Starts the default service
Start Windows Service	svcstart.bat	Starts the default service as a Windows service
Stop Windows Service	svcstop.bat	Stops the default service, if it was started as a Windows service
Uninstall Windows Service	remove.bat	Uninstalls the default service, if it was installed as a Windows service

The file **appstart.sas**, located in the Service Directory subfolder, is used to start PROC APPSRV and, as we shall see shortly, to allocate user-defined libraries and catalogs on the server. (This file is identical to that on the UNIX platform.)

Creating a Windows service takes advantage of the ability of Windows NT and subsequent versions (Windows 2000 and XP) to initiate a background service at startup. (On UNIX platforms, the service must be started manually or added to the **init.d** configuration.) If you select **Install as Windows Service**, the default service will be started every time the system is booted. If you change your mind about having the service start automatically, SAS has provided a useful batch script to uninstall the Windows service.

If you choose **Start Windows Service** or **Start Interactively** (depending on whether you installed the Application Dispatcher `default Service` as a Windows service as discussed previously), you should be able to run the examples that SAS has provided as shown in the following section.

Currently, the Application Dispatcher can be used to stop the service on both UNIX and Windows, but not to start it. Just send the following URL to the server to stop the service:

```
http://<server-name>/cgi-bin/broker?service=default&program=stop
```

The following message should appear in the browser window:

```
The Application Server <server-name>.<domain-name>:5001 has
been stopped.
```

Testing the Application Service

After you have installed the Application Broker and defined and started the default application service, it's finally time to have some fun and try it out. Remember that the Application Broker is an executable file in the **cgi-bin** directory; on Windows it is called **broker.exe**, and on UNIX just **broker**. Since it is a CGI applications program, we can call it with one of the following URLs:

- Apache – `http://<server-name>/cgi-bin/broker?`
  `_service=default&_program=status`

- IIS – `http://<server-name>/scripts/broker.exe?`
  `_service=default&_program=status`

Display 6.8 shows the result of running the Application Broker on a Windows platform Web server with two name/value pairs as parameters: `_service` (the service name) and `_program` (the program name, in this case `status`). This is a built-in utility program supplied with the Application Dispatcher software.

Display 6.8 Testing the Application Broker on Windows

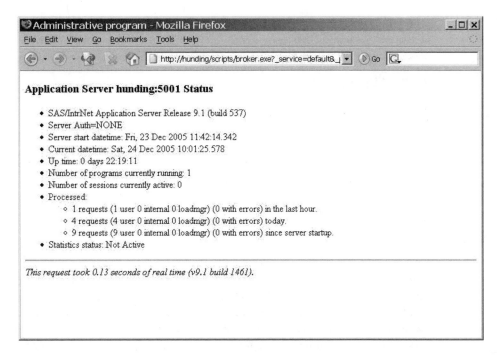

Remember the quote from the documentation back at the beginning of the chapter that said in essence that all you need is a service name and a program name? This is what they were talking about. The parameters are appended to the URL just as we saw in the Perl program in Chapter 5. The Application Broker passes a request for SAS program code to the Application Server, which locates the program, executes it, and sends the result back to the broker. (Actually, the _service parameter is not required if the DefaultService directive is set in the **broker.cfg** file and you want to use that service.)

Changing the value of the _program parameter runs a different program. For example, Display 6.9 shows another familiar-looking Web page. As the display illustrates, the URL for this page (all in one line in the browser) is as follows:

```
http://hunding/scripts/broker.exe?
    _service=default&
    _program=sample.webhello.sas
```

Note that on Windows the URL is not case-sensitive, while on a UNIX system it would be.

Display 6.9 SAS/IntrNet Samples: Hello World!

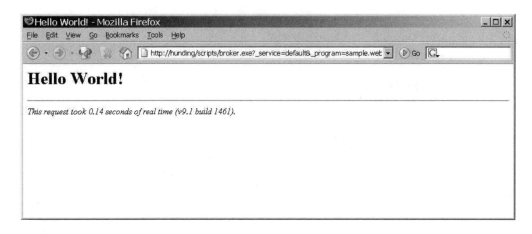

In this example the URL includes a new value for the `_program` parameter: `sample.webhello.sas`. The `webhello.sas` sample program, along with a number of others, is supplied with the SAS/IntrNet software. On UNIX servers these are installed by default in the directory **$SASROOT\samples\IntrNet**; on Windows the directory is **$SASROOT\IntrNet\sample** (go figure). The program name must include three levels: `sample` is the name of the program library (not the directory; see Example 6.1), while `webhello.sas` is the filename of the program itself. The program code is shown in Example 6.1:

Example 6.1 SAS/IntrNet: Hello World

```
/*simply write out a web page that says "Hello World!"*/
data _null_;
    file _webout;
    put '<HTML>';
    put '<HEAD><TITLE>Hello World!</TITLE></HEAD>';
    put '<BODY>';
    put '<H1>Hello World!</H1>';
    put '</BODY>';
    put '</HTML>';
run;
```

This simple program illustrates how easy it is to create HTML using the Application Dispatcher. The most important line in the DATA step is the highlighted one, defining the file reference for the output as `_webout`. This is a SAS FILEREF that is automatically defined for you when the Application Server starts, indicating that the output of the PUT statements should be sent to the Web client. SAS/IntrNet software automatically takes care of inserting the correct mime-type header and sending the code back to the client browser.

Of course, usually one needs to send more parameters to the server than just the name of the program. The Application Dispatcher simply translates all the name/value pairs into macro variables that are passed to the SAS program; this is the SAS equivalent to the Perl `parameter()` function described in the preceding chapter. In fact, we can recreate the temperature conversion program easily in SAS as follows:

Example 6.2 SAS/IntrNet: Temperature Conversion Program

```
/* SAS/IntrNet program to convert F to C and vice versa */

data _null_;

 file _webout;

 ***** write generic XHTML header *****;
 put '<?xml version="1.0" encoding="utf-8"?>'/
   '<!DOCTYPE html PUBLIC ""-//W3C//DTD XHTML Basic 1.0//EN"'/
   ' "http://www.w3.org/TR/xhtml-basic/xhtml-basic10.dtd">'/
   '<html xmlns="http://www.w3.org/1999/xhtml" lang="en-US">';

 ***** write top of page *****;
 put '<head>'/
   '<title>SAS/IntrNet Temperature Conversion
   Calculator</title>'/
   '</head>'/
   '<body>'/
   '<div style="text-align: center">'/
   '<h1 style="color: blue">Temperature Conversion
   Calculator</h1>';

     **** create HTML form * with hidden text fields ***;
     put '<form name="calculator" action="' &_URL
     '" method="get">'/
       '<input type="hidden" name="_service" value="default" />'/
       '<input type="hidden" name="_program" value="' &_program
       '" />';

 ***** get parameter values *****;
 temp = symget('input');
 type = symget('convert');

 ***** input temperature *****;
 put '<p><strong>Enter a temperature and select a conversion
     type: </strong>'/
     '<input type="text" name="input" value="' temp '" /></p>';

 ***** select conversion/compute result *****;
 put '<p><input type="radio" name="convert" value="1"' @;
 if (type = '1') then do;
     put 'checked="checked"' @;
     result = 5*(input(temp,8.) - 32)/9;
 end;
 put '> Fahrenheit to Centigrade</p>';
```

```
put '<p><input type="radio" name="convert" value="2"' @;
if (type = '2') then do;
    put 'checked="checked"' @;
    result = 9*input(temp,8.)/5 + 32;
end;
put '> Centigrade to Fahrenheit</p>';

***** Submit button *****;
    put '<p><input type="submit" value = "Submit"></p>';

***** Display results *****;
if (temp > ' ') then do;
    put '<p><strong>Result: </strong>'/
        '<input type="text" name="result" value="' result 6.2
        '"/></p>';
end;

**** write bottom of page *****;
put '</form>'/
    '</div>'/
    '</body>'/
    '</html>';
run;
```

Comparing this program to the Perl version in the previous chapter, we can see a number of evident differences:

- SAS includes the mime-type header automatically; in Perl the `header` function is required.

- Perl has `start_html` and `end_html` functions that simplify HTML housekeeping chores, while in SAS these must be done explicitly.

- SAS requires the `_webout` file destination to send output back to the Dispatcher.

- SAS uses special reserved macro variables for parameters, shown in bold: `_program` is the name of the Dispatcher program that the Application Server should run, while `_url` is a reference to the Application Broker CGI program (see http://support.sas.com/rnd/web/intrnet/dispatch/reffield.html for a list of the reserved variables available in the Application Dispatcher).

Overall, however, the Perl and SAS programs have the same effect, as shown in Display 6.10, where the URL points to the code in Example 6.2:

```
http://hunding/scripts/broker.exe?
    _service=default&
    _program=sample.example6-2.sas
```

Display 6.10 SAS/IntrNet: Temperature Conversion Output

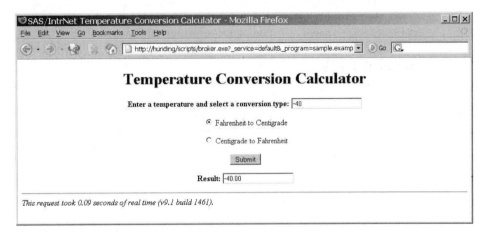

The first time the page is called, the URL needs only the required `_service` and `_program` parameters. The user then fills in the input box, checks a conversion type, and clicks the **Submit** button. Pressing **Submit** causes the Web server to send the values of all the input fields (including the hidden ones) to the Application Dispatcher. (Blanks in a character variable value are changed to plus signs (+) by the server; SAS text formatting functions such as `inputn`, `left`, `trim` or `strip` can be used to specify the desired number of decimal places.)

There are five name/value pairs in this program: `_service` and `_program` parameters are passed to the broker to identify the service and program. The `input` and `convert` fields are used to collect the user input and pass it to the SAS program as macro variable values. Finally, the `result` field is used to display the answer; this field is also passed as a parameter that is never read by the SAS program.

Defining Application Server Libraries

One tricky point that was glossed over in the previous example is the value of the `_program` parameter: in this case, `sample.example6-2.sas`. The question of where to store the SAS program and how to reference it brings us to the promised discussion of editing the **appstart.sas** program. You probably do not want to store your code in the same directory as the samples supplied by SAS, so how do you define a new library to the service?

Recall that the batch scripts used to start the default SAS application service are in a subfolder called **default** in the root directory defined by the inetcfg utility (see Display 6.2). These scripts are used to run the **appstart.sas** program which contains the APPSRV procedure along with the various program statements that define the server.

A default version of this program is created by the `inetcfg` utility; in this case, the path to the program is **`C:\Program Files\SAS\IntrNet\default\appstart.sas`**. The ALLOCATE statement, shown in bold, associates the filename **`examples`** with the specified directory. The PROGLIBS statement indicates that the referenced directory contains SAS programs.

Example 6.3 PROC APPSRV

```
proc appsrv  unsafe='&";%''' &sysparm;
  allocate file sample '!SASROOT\IntrNet\sample';
  allocate library samplib '!SASROOT\IntrNet\sample'
  access=readonly;
  allocate library sampdat '!SASROOT\IntrNet\sample'
  access=readonly;
  allocate library tmplib 'C:\Program
  Files\SAS\IntrNet\default\temp';
  allocate file logfile
  'C:\Program Files\SAS\IntrNet\default\logs\%a_%p.log';

  /* allocate new program library */
  allocate file examples 'c:\Inetpub\scripts\examples';
  proglibs examples;

  proglibs sample samplib %ifcexist(sashelp.webeis)
  sashelp.webprog;
  proglibs sashelp.websdk1;
  adminlibs sashelp.webadmn;
  datalibs sampdat tmplib;
  log file=logfile;
```

In this way, the library **`C:\Inetpub\scripts\examples`** can be referenced in a URL as **`examples`**. Note that it is generally considered not a good idea to put your SAS programs in the directory reserved for Web server scripts. This is akin to putting your personal SAS programs in a directory like **`C:\Program Files\SAS\SAS 9.1\intrnet\sashelp`**. Installing or changing the SAS installation could result in a loss of your files. Similarly, changing the installation of the Web server software could result in the loss of your files. And storing the files together adds great confusion, because later it is stated that unlike CGI programs, the SAS programs don't need to be stored in a directory that the Web server recognizes as a CGI alias. A better choice might to define a new folder such as **`C:\examples`** to hold the SAS code.

In any case, if the file **`webhello.sas`** is located in the folder specified, the URL (all on one line) would now be as follows:

```
http://hunding/scripts/broker.exe?_service=default&
  _program=examples.webhello.sas
```

There are several kinds of Application Server libraries; the most important are *program libraries* and *data libraries*. Two SAS statements are required to define each:

- Program libraries allocate a file reference to a directory and then include the FILENAME reference in a PROGLIBS statement; only programs that reside in a library designated by a PROGLIB statement can be executed.

- Data libraries allocate a library reference to a directory and then include the LIBNAME reference in a DATALIBS statement; DATALIB libraries are intended to store data, not programs.

You can also use the ALLOCATE LIBRARY statement to allocate program libraries for executing SCL, SOURCE, or MACRO catalog entries. You can include as many ALLOCATE, PROGLIBS and DATALIBS statements as you wish.

The Application Broker program is a CGI program, so it must be stored in a directory that the Web server recognizes as a CGI alias. In contrast, you can store the SAS programs anywhere you want; just be sure to modify **appstart.sas** by adding the appropriate directory references.

A valid system administrator account must be used to modify the **appstart.sas** program. It is also important to note that you must stop the Windows service before modifying the program, or else you will get a "Sharing Violation" message when you try to save it. Just don't forget to restart the service after saving the changes.

Obviously, there are a number of other options available in this procedure; the developer is directed to the APPSRV procedure syntax in the SAS Help and Documentation for more information about the other statements shown. (See also http://support.sas.com/rnd/web/intrnet/dispatch/appsrv.html).

Debugging SAS Output

The default HTML message when a CGI program fails for some reason is that the program has an error. This can be somewhat frustrating, since there is no way to figure out what is causing the error. SAS/IntrNet has a handy device for displaying the input parameters and the log file of the SAS program. If nothing else, this would justify using SAS for CGI programming. Just add the debug flag to the end of the URL string, as in the following example:

```
http://hunding/scripts/broker.exe?
  _service=default&
  _program=examples.webhello.sas&
  _debug=131
```

The numeric values of the debug flag for the Application Dispatcher are computed as the sum of the codes in Table 6.2; thus 131=1+2+128.

Table 6.2 List of Valid Debug Values

Value	Keyword	Description
1	FIELDS	Echoes all fields. This is useful for debugging value-splitting problems.
2	TIME	Prints the broker version number and elapsed time after each run—for example, "This request took 2.46 seconds of real time (v8.0 build 1316)." Also, this value displays the Powered by SAS logo if you provide additional settings as described in "Displaying the Powered by SAS Logo" in the SAS Help and Documentation.
4	SERVICES	Lists definitions of all services as defined by the administrator, but does not run the program.
8		Skips all execution processing.
16	DUMP	Displays output in hexadecimal. This is extremely helpful for debugging problems with the HTTP header or graphics.
32		Displays the Powered by SAS logo without broker version or elapsed time information. See also "Displaying the Powered by SAS Logo" in the SAS Help and Documentation.
128	LOG	Returns the log file. This is useful for diagnosing problems in the SAS code.
256		Traces socket connection attempts. This is helpful for diagnosing the machine selection process.
512		Shows the socket host and port number in the status message (by default this is off for security reasons).
1024	ECHO	Echoes data usually sent from the broker to the Application Server. It does not run the program. In the case of a launch service, this also shows the SAS command that would have been invoked by the broker.
2048	TRACE	Provides more extensive socket diagnostics.
4096		Prevents the deletion of temporary files that are created for launch. This is useful for debugging configuration problems in a launch service (prior to SAS 8).
8192		Returns the entire SAS log file from a launched service (prior to SAS 8).
16384	ENV	Displays the broker environment parameters.

Note that the use of the keywords is not supported in the SAS 8 broker; that support was added to the broker in SAS®9.

Display 6.11 illustrates the results of choosing the debug values for echoing fields (1), broker version (2), and the log (128). If there were any errors in the execution of the SAS code, they would show up in the log.

Display 6.11 SAS/IntrNet: Debug Flags

```
Symbols passed to SAS

#symbols: 20
 "_SRVNAME" = "hunding"
 "_SRVPORT" = "80"
 "_REQMETH" = "GET"
 "_RMTHOST" = "192.168.2.101"
 "_RMTADDR" = "192.168.2.101"
 "_RMTUSER" = ""
 "_HTCOOK" = ""
 "_HTUA" = "Mozilla/5.0 (Windows; U; Windows NT 5.1; en-US;
  rv:1.7.12) Gecko/20050915 Firefox/1.0.7"
 "_mrvimg" = "/sasweb/IntrNet9/MRV/images"
 "_grfaplt" = "/sasweb/graph/graphapp.jar"
 "_grafloc" = "/sasweb/graph"
 "_service" = "default"
 "_program" = "examples.webhello.sas"
 "_debug" = "131"
 "_VERSION" = "9.1"
 "_URL" = "/scripts/broker.exe"
 "_ADMIN" = "Your Name"
 "_ADMAIL" = "yourname@yoursite"
 "_SERVER" = "hunding"
 "_PORT" = "5001"

Using timeout: 60
Content-type: text/html Pragma: no-cache Content-type:
text/html
Hello World!


Sas Log for this Request

NOTE: running request program examples.webhello.sas

NOTE: %INCLUDE (level 1) file
c:\Inetpub\scripts\examples\webhello.sas is file
c:\Inetpub\scripts\examples\webhello.sas.

2  + /*******************************************************/
3  + /*         S A S   S A M P L E   L I B R A R Y         */
4  + /*                                                     */
5  + /*     NAME: WEBHELLO                                  */
6  + /*     TITLE: Hello World                              */
7  + /*   PRODUCT: SAS/IntrNet (Application Dispatcher)     */
8  + /*    SYSTEM: ALL                                      */
9  + /*      KEYS:                                          */
10 + /*     PROCS:                                          */
11 + /*      DATA:                                          */
12 + /*                                                     */
13 + /*   SUPPORT: Web Tools Group        UPDATE: 13Oct2000 */
14 + /* REF: http://support.sas.com/rnd/web/IntrNet/dispatch/*/
15 + /*      MISC:                                          */
16 +/*******************************************************/
17 +
18 + /*simply write out a web page that says "Hello World!"  */
19 +data _null_;
```

```
20 +  file _webout;
21 +  put '<HTML>';
22 +  put '<HEAD><TITLE>Hello World!</TITLE></HEAD>';
23 +  put '<BODY>';
24 +  put '<H1>Hello World!</H1>';
25 +  put '</BODY>';
26 +  put '</HTML>';
27 +run;

NOTE: The file _WEBOUT is:
      Access Method=Application Server Access Method,
      Network connection type=Full Duplex,
      Peer IP address=192.168.2.102

NOTE: 6 records were written to the file _WEBOUT.
      The minimum record length was 6.
      The maximum record length was 40.
NOTE: DATA statement used (Total process time):
      real time            0.00 seconds
      cpu time             0.00 seconds

NOTE: %INCLUDE (level 1) ending.
NOTE: request has completed

This request took 0.11 seconds of real time (v9.1 build 1461).
```

The SAS Application Dispatcher documentation suggests setting some of these flags off in production environments so that users will not have access to the information. This is accomplished by using the `DebugMask` and `ServiceDebugMask` directives in the file **broker.cfg** (see http://support.sas.com/rnd/web/intrnet/dispatch/debuginp.html).

Generating Dynamic Output with the Output Delivery System

Most users are going to want to use SAS procedures to display output, rather than PUT statements. There are two ways to accomplish this: using the old SAS 6 Web Publishing Tools or the Output Delivery System. (See "The Program Component" at http://support.sas.com/rnd/web/Intrnet/dispatch/procomp.html for more detail on writing SAS programs for the Application Dispatcher.)

As the discussion in Chapter 3, "Creating Static HTML Output," suggests, ODS is the preferred approach. The following simple example shows how to create a Web page containing a table based on the current values stored in a SAS data set; the example is simplified from the Application Dispatcher samples available with the default SAS/IntrNet installation. (You can get more information on this and the other Application Dispatcher examples at http://support.sas.com/rnd/web/intrnet/dispatch/sampapp.html.)

The output of the REPORT procedure is directed to the Application Dispatcher using ODS, which sends the HTML output, created by the HTML3 destination, to the file reference _webout.

Example 6.4 Using ODS to Display Procedure Output

```
%* Sales report Example - Display Product by Region;
%macro salesrpt;

 %global region;
  proc report data=sashelp.shoes;
     by region;
     %if (&region ne ) %then %do;
        where region="&region";
     %end;
     title "Sales by Product";
     footnote "Data are current as of &systime &sysdate9";
     column product sales;
     define product / group;
     define sales / analysis sum;
   quit;

%mend salesrpt;

/* redirect output to client */
ods html3 body=_webout;

%salesrpt
```

There are a couple of simple tricks in this program. The SAS macro variables SYSDATE9 and SYSTIME are used to display the date and time the program is run. In order to ensure that the program does not fail if the user leaves off the REGION parameter, the macro variable is initialized with a %GLOBAL statement; if no region is specified, the default is to display all of the data.

The URL for this page illustrates the three name/value pairs required to call the sample program for the region "Africa"; the output is shown in Display 6.12.

```
http://hunding/scripts/broker.exe?
   _service=default&_program=examples.example6-4.sas&
   region=Africa
```

Note that this example uses the html3 output destination. The SAS Notes indicate that the Application Server may crash if you use ODS and do not properly close the listing; see http://support.sas.com/techsup/unotes/SN/011/011595.html.

Display 6.12 SAS/IntrNet: Directing Procedure Output to the Web

You can also use the ODS MARKUP statement to specify a user-defined template, as shown in Chapter 3. In that case, include something like the following code instead of the ODS HTML3 statement:

Example 6.5 Using ODS MARKUP to Customize Procedure Output

```
/* prepend XHTML template */
libname userlib c:\Documents and Settings\frederick\
   My Documents\My SAS Files\9.1
ods path (prepend) userlib.templat;

/* redirect output to client */
ods markup body=_webout tagset=xhtml;
```

In this case, the XHTML template illustrated in Example 3.19 has been loaded into the SASUSER.TEMPLAT item store, as explained in Chapter 3. The template is then prepended to the ODS markup statement, so that the custom tagset will be used to format the resulting XHTML.

In practice, the page would almost certainly be called from an HTML form with a list of the available regions in order to make sure that the values are spelled correctly. In order to get the list, you could use the Application Dispatcher again, or you could choose to run a query against the SAS data set using htmSQL. How to do this is the topic of the next chapter.

References

SAS Documentation
URL references are current as of the date of publication.

- Application Server Debugging –
  http://support.sas.com/rnd/web/intrnet/dispatch/debuginp.html
- Configure Application Server –
  http://support.sas.com/rnd/web/intrnet/dispatch/appsrv.html
- Configure Broker – http://support.sas.com/rnd/web/intrnet/dispatch/bconfig.html
- Create Default Service –
  http://support.sas.com/rnd/web/intrnet/dispatch/defsvc.html
- Programming with the Application Dispatcher –
  http://support.sas.com/rnd/web/intrnet/dispatch/procomp.html
- Sample Applications –
  http://support.sas.com/rnd/web/intrnet/dispatch/sampapp.html
- Using ODS with the Application Dispatcher –
  http://support.sas.com/rnd/web/intrnet/dispatch/ods.html

SAS Publications

- Buffington, Stacy, and Doyle McDonald. 1999. "Collecting Data Via the Internet with SAS/IntrNet and SAS/SHARE Software." *Proceedings of the Twenty-Fourth Annual SAS Users Group International Conference.* Cary, NC: SAS Institute Inc.
- Rafiee, Dana. 2001. "Dynamic Web-Based Applications Using SAS/IntrNet Software." *Proceedings of the Twenty-Sixth Annual SAS Users Group International Conference.* Cary, NC: SAS Institute Inc.
- Rose, Roderick A. 2001. "The Basics of Dynamic SAS/IntrNet Applications." *Proceedings of the Twenty-Sixth Annual SAS Users Group International Conference.* Cary, NC: SAS Institute Inc.
- Rosenbaum, Andrew. 2002. "A Look at the Development Process for a SAS/IntrNet Application." *Proceedings of the Twenty-Seventh Annual SAS Users Group International Conference.* Cary, NC: SAS Institute Inc.
- Slaughter, Susan J., Sy Truong, and Lora D. Delwiche. 2002. "ODS Meets SAS/IntrNet." *Proceedings of the Twenty-Seventh Annual SAS Users Group International Conference.* Cary, NC: SAS Institute Inc.
- Timbers, Vincent. 2002. "Connecting the SAS System to the Web: A Hands-on Introduction to SAS/IntrNet Application Dispatcher." *Proceedings of the Twenty-Seventh Annual SAS Users Group International Conference.* Cary, NC: SAS Institute Inc.
- Ullah, Ahsan. 1999. "Customize your WEB output using SAS/IntrNet Software and WEB Publishing Tools." *Proceedings of the Twenty-Fourth Annual SAS Users Group International Conference.* Cary, NC: SAS Institute Inc.

Chapter 7

SAS/IntrNet: htmSQL

SAS/IntrNet Data Services

One of the most common (and most useful) Web programming tasks is generating dynamic Web pages based on database queries. In addition to the Application Dispatcher, which provides server-side computing services, SAS/IntrNet also offers the htmSQL query tool. This CGI-based utility is the simplest and most powerful database access tool currently available. This chapter offers some easy htmSQL examples that suggest the power and flexibility of this tool.

In order to use htmSQL, it is essential to be familiar with the syntax of the SAS SQL procedure. SQL, the Structured Query Language, is the industry standard for access to relational databases. SAS has provided a super-set of this standard that includes a query language for SAS data libraries. There are many good tutorials available on SQL in general and on the SAS implementation in particular; the reader is referred to the many good online resources, printed books, and manuals.[1]

---

[1] For a list of several available SQL tutorials, see http://dir.yahoo.com/Computers_and_Internet/Programming_and_Development/Languages/SQL/Tutorials/.

In addition to knowledge of PROC SQL syntax, htmSQL also requires some familiarity with setting up a SAS/SHARE server. (See Chapter 4, "Remote Access to SAS," for an introduction to using SAS/SHARE.) In the examples that follow, the SAS data library referenced is located on a UNIX server that is connected to the Web server via a TCP/IP connection. Remember that all database systems, including SAS, require a database engine running as a background process. In this case, the SAS/SHARE server is on a separate platform, with a different operating system, from the Web server. Using the magic of TCP/IP, SAS/IntrNet can run PROC SQL queries against this database engine and display the results in the client Web browser.

htmSQL Directives

The htmSQL tool, like the Application Broker, is a CGI program located on the Web server. (There is an `htmSQL.cfg` file, analogous to the `broker.cfg` file, but it is unlikely that you will ever need to modify the default values installed.) You invoke htmSQL by specifying a URL such as one of the following:

- UNIX/Linux: http://*<server-name>*/cgi-bin/htmSQL/sample.hsql
- Windows IIS: http://*<server-name>*/scripts/htmsql.exe/sample.hsql

Note that although the htmSQL tool is located in the `cgi-bin` (or `scripts`) directory, the Web page itself is stored with an extension of `.hsql` in the regular document root directory. The default file extension for a Web page containing htmSQL directives is not `.html` but `.hsql`. In comparison to the Application Broker, which requires the `_service` and `_program` parameters, htmSQL simply opens the `.hsql` file from the specified location. In other words, unlike the Application Broker, you do not have to make any modifications to htmSQL to get it to run. Just create a Web document, store it in a public HTML directory, and reference it as shown in the preceding example. (Note also that you should not put a question mark (?) before the name of the `.hsql` page.)

The big difference between an HTML page and an `.hsql` page is that the latter includes commands that process SQL statements and the resulting queries. These commands are referred to as htmSQL *directives*. There is a short list of directives to learn, although the specific syntax can get a little complicated. (See http://support.sas.com/rnd/web/intrnet/htmSQL/syntax.html for a complete list of directives.)

All directives require an opening and closing tag for each, enclosed in { } brackets. The following four are most often used:

- {query} defines the name of the SAS/SHARE server session; all other htmSQL directives must be contained between {query} and {/query} directives.

- {Sql} contains the SQL query.

- {Eachrow} contains HTML tags to format the query results.

- {Norows} contains HTML text that is displayed if no data is returned by the query.

Within the query section, *variable reference directives* are used to specify the values to be inserted. When htmSQL encounters a variable reference within an eachrow section, it replaces the tag with the current row value of the named variable or variables. If the reference is before the eachrow section, the values are that of the first row; if after, then the values are that of the last row.

Variable reference directives are of the form {&varname[value]}. Although these may look like macro references, do not confuse the two. Note also that unlike the other htmSQL directives, variable reference directives have no closing tags. The three possible formats for variable reference directives are shown in Table 7.1:

Table 7.1 htmSQL Variable Reference Syntax

Directive	Reference
{&varname[n]}	the *n*th value that the variable contains, where n >= 1
{&varname[m..n]}	all values from the *m*th value to the *n*th value where m <= 1 < n and the variable contains two or more values
{&varname[*]}	all values that the variable contains

For example, in the following SQL statement,

```
{sql}
   select fname, mname, lname from employee.names
{/sql}
```

the values of the {&varname} directive would be resolved as follows:

- {&varname[1]}   = fname

- {&varname[2..3]} = mname lname

- {&varname[*]}   = fname mname lname

The most important optional arguments on a variable reference directive are BEFORE=, BETWEEN= and AFTER=; these specify the HTML formatting tags to appear before, between and after the variable values. htmSQL also defines a number of *automatic* variable references, including date and time variables and SQL-related variables. These have the prefix sys, as in the following examples. The use of date-time and SQL-related variables is also illustrated in Example 7.1. (A complete list of the automatic variables is available at http://support.sas.com/rnd/web/intrnet/htmSQL/autovar.html.)

- `sys.time` – the current time of day, based on the Web server system clock
- `sys.colname[value]` – the name of one or more columns in the query, where `[value]` can be one of the three choices shown previously for the `{&varname}` directive
- `sys.qrow` – the number of the current row

Generating Tables with htmSQL

In general, the Application Dispatcher can be used to display procedure output and htmSQL the results of SQL queries. As any good PROC SQL programmer knows, however, you can do an awful lot with SQL. The following example replicates the results of Example 6.4 in the previous chapter, using SQL statements instead of PROC REPORT.

Example 7.1 Summarizing Data with htmSQL

```
<!DOCTYPE html
    PUBLIC "-//W3C//DTD XHTML Basic 1.0//EN"
    "http://www.w3.org/TR/xhtml-basic/xhtml-basic10.dtd">
<html xmlns="http://www.w3.org/1999/xhtml" lang="en-US">

<head>
    <title>SAS IntrNet Examples: htmSQL</title>
    <style type="text/css">
        body    { text-align: center; }
        caption { font-weight: bold; }
        h1      { color: blue; }
        h3      { color: red; }

        td      { text-align: right; }
        p.foot  { font-weight: bold; }
    </style>
</head>

<body>
{query server="hygelac:5011"
  sapw="user" userid="sas" password="sasuser" }
{sql}
  select product,
      sum(sales) as total label="Total Sales" format=dollar8.
  from sashelp.shoes
  where region='{&region}'
  group by product
{/sql}
<h1>Shoe Sales by Product</h1>
<hr/>
```

```
{norows}
   <h3>No rows selected. Check that the region parameter
   has been specified correctly.</h3>
{/norows}
<table border="1" cellpadding="4" cellspacing="0"
      align="center">
   <caption>Region: {&region}</caption>
   <tr>
      {label var="{&sys.colname[*]}"
                  before="<th>"
                  between="</th><th>"
                  after="</th>"}
   </tr>
   {eachrow}
      <tr>
      { &{&sys.colname[*]}
            before="<td>"
            between="</td><td>"
            after="</td>" }
      </tr>
   {/eachrow}
</table>
{/query}
<hr/>
<p class="foot">Data are current as of {&sys.time}{&sys.ampm}
   {&sys.month} {&sys.monthday}, {&sys.year}
</p>
</body>
</html>
```

The output of this program is very close to that produced using the Application Dispatcher with PROC REPORT; the difference is that the summarization is done in SQL and the formatting with HTML. The URL for this page is

```
http://hunding/scripts/htmSQL.exe/example7-1.hsql?region=Africa
```

The Web server in this case is IIS on the Windows platform, while the SAS/SHARE server is on a separate computer that is running the Linux version of the UNIX operating system.

The htmSQL application expects to find the document source in an Apache directory, not in **scripts** or **cgi-bin**. Remember, the htmSQL utility is located in the **cgi-bin** directory; in this example **C:\Inetpub\scripts** on the Web server, while the source file **Example7-1.hsql** is in the document root directory, **C:\Inetpub\wwwroot**.

Display 7.1 htmSQL Table Output

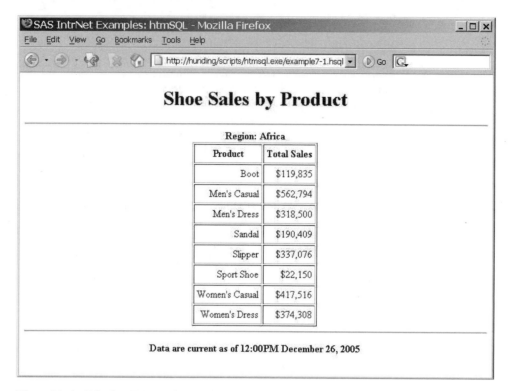

The table in Display 7.1 can be produced by the following steps:

1. An internal CSS stylesheet is included in the page to format the resulting page.

2. The htmSQL {query} directive opens a SAS/SHARE session on the UNIX server. Note that for this example to work, the SAS/SHARE server must be started on port 5011 of the remote Linux host, so that the htmSQL query will find the server.

3. The {sql} directive embedded within the {query} directive submits the SELECT query to the specified SAS data set on the SAS/SHARE server.

4. User-supplied parameter values are specified by supplying a macro variable name in brackets (e.g., {&region}).

5. The {norows} directive specifies the text to be displayed if the query does not return any data (for example, if a REGION parameter is not specified).

6. An HTML table is created to display the results of the query (see Chapter 3, "Creating Static HTML Output," for a review of how to create tables).

7. The HTML table header row is generated, using the label parameter on the SQL-related variable reference directive {&sys.colname[*]}.

8. The {eachrow} directive specifies a loop over the SQL row set. The table rows are generated using the double ampersand form of the SQL-related variable reference directive &{&sys.colname[*]}. htmSQL interprets this as an abbreviation for all the fields in the query row.

9. Finally, the footnote is produced by a series of date-time reference variables.

Clearly, htmSQL can be used in place of the Application Dispatcher for any task that can be coded using the SQL procedure. The documentation indicates that it is likely to be faster for simple lists or tables. Furthermore, SQL is an easy and natural way to create HTML forms, as we shall see in the following section.

The main disadvantages of htmSQL are, first, that it requires licensing and configuring an additional product, the SAS/SHARE server; and second, that it lacks the flexibility and power of the full SAS programming language for data management and analysis.

Creating Forms with htmSQL

The big problem with running the query in Example 7.1 is that you have to specify a valid value for the REGION parameter. Clearly, this would be a great opportunity for an HTML form that allows the user to select a region from a list of values in the database. With htmSQL, nothing (well, almost nothing) could be simpler. The Web page shown in Display 7.2 illustrates a sample form as it would appear with the **Select Region** drop-down list open:

Display 7.2 Using htmSQL to Generate HTML Form Controls

Clicking the **Submit Query** button sends the following URL to the Web server:

```
http://hunding/scripts/htmSQL.exe/example7-1.hsql?
   region=United+States
```

The result is the report of sales by product for the selected region, shown in Display 7.1.

The .hsql source code for Example 7.2 is similar to the other htmSQL we have seen; directives are shown in bold. The SQL SELECT DISTINCT statement creates a list of unique REGION values. The `{&region}` variable reference in the eachrow section creates the option box displayed in Display 7.2. That's all there is to it!

Example 7.2 htmSQL Code to Generate Form Control Values

```
<!DOCTYPE html PUBLIC "-//W3C//DTD XHTML Basic 1.0//EN"
   "http://www.w3.org/TR/xhtml-basic/xhtml-basic10.dtd">
<html xmlns="http://www.w3.org/1999/xhtml" lang="en-US">
<head>
   <title>SAS IntrNet Examples: htmSQL</title>
</head>
<body>
  {query server="hygelac:5011"
      sapw="user" userid="sas" password="sasuser" }
  {sql} select distinct region from sashelp.shoes{/sql}
  <div style="text-align: center; ">
  <form method=get
action="http://hunding/scripts/htmSQL.exe/example7-1.hsql">
      <h1 style="color: blue;">International Shoe Company</h1>
      <h2>Request Current Sales Report</h2>
      <p>Select Region:
        <select name=region>
            {eachrow}<option>{&region}</option>{/eachrow}
        </select>
        <input type="submit" />
      </p>
  </form>
  </div>
  {/query}
</body>
</html>
```

Issues with CGI Applications

CGI-based programs can be powerful and flexible, but they do come with some drawbacks. Primary among these is that CGI technology can be resource intensive, since each new request to the server starts a new operating system process. It should be noted, however, that this constraint largely does not apply to the Application Broker.

The broker is written in C and operates as a transient "traffic cop," directing traffic between the client and the SAS server. The SAS server does all the heavy lifting. Most other CGI applications don't have the luxury of having a SAS server available, and thus must do all the heavy lifting themselves. Yes, other CGI applications can open connections to database servers, but once they get the data, THEY need to crunch/format/present it. In contrast, the SAS server does all the heavy processing, and the broker just passes the output back to the client with little overhead. In the 11 years that SAS/IntrNet has been in the field, we have had few, if any, reported performance problems that were due to the broker consuming too many resources and choking. This cannot be said for other traditional CGI applications, which are typically written using an interpreted language such as Perl.[2]

---

[2] Quote from SAS developer Vince DelGobbo, February 15, 2006, used with permission.

There are also some unavoidable security issues with CGI. Consequently, the newest technologies for server-side Web development, based on the Java servlet API or Microsoft Active Data Objects, are gradually replacing CGI applications in many sites. This doesn't mean that there is no point in using the SAS/IntrNet CGI tools, but only that it is important to be aware that there are alternatives. The next section of this text covers using SAS AppDev Studio for creating server-side Java applications.

References

SAS Documentation
URL references are current as of the date of publication.

- SAS Institute, Inc. 1995. *Getting Started with the SQL Procedure.* 1st ed. Cary, NC: SAS Institute Inc.

- SAS Institute, Inc. 2000. *SAS SQL Procedure User's Guide, Version 8.* Cary, NC: SAS Institute Inc.

- SAS Institute, Inc. 2001. *SQL Processing with the SAS System* (Course Notes). Cary, NC: SAS Institute Inc.

- SAS Institute, Inc. Automatic Variables – http://support.sas.com/rnd/web/intrnet/htmSQL/autovar.html

- SAS Institute, Inc. Directives – http://support.sas.com/rnd/web/intrnet/htmSQL/syntax.html

SAS Publications

- Wyland, Karen L. 2002. "Ask DSS…using HTML, JavaScript, htmSQL, and SAS to Create Dynamic Web Applications." *Proceedings of the Twenty-Seventh Annual SAS Users Group International Conference.* Cary, NC: SAS Institute Inc.

Part 3

Server-Side Java Programming

Chapter **8**

Java Servlets and JavaServer Pages

Introduction

The perceived limitations of CGI (see Chapter 7, "SAS IntrNet: htmSQL,") have led to a number of alternative strategies for delivering dynamic content on the Web. Of these, the most commonly used are Microsoft Active Server Pages (ASP), PHP (Linux freeware) and JavaServer Pages (JSP) from Sun. These protocols do not rely exclusively on scripts that run within HTML forms (like CGI and DHTML), but instead use a more sophisticated object-oriented approach. Active Server Pages use *Active Data Objects (ADO)* to provide a distributed information environment for Microsoft operating systems. The equivalent in the cross-platform world is Sun's *J2EE Web Technology* including servlets, JavaServer Pages and JavaBeans.

SAS AppDev Studio provides comprehensive support for cross-platform development. Chapter 9, "Developing Java Server-Side Applications with webAF Software," illustrates how to use webAF software to implement server-side access to SAS data sets, while Chapter 10, "Using the SAS Open Metadata Architecture with the Integrated Object Model," introduces SAS Integration Technologies and the SAS Open Metadata Architecture. This chapter is a short introduction to the major concepts of Java servlets and JavaServer Pages. For more detail and examples of how to implement JSP, see the Sun online documentation at http://java.sun.com/products/jsp/. The syntax reference for JSP v1.2 is at http://java.sun.com/products/jsp/syntax/1.2/syntaxref12.html. This chapter and the following one assume some familiarity with Java programming, classes, and methods. JSP is a large and complex set of concepts, and the following exposition assumes an understanding of the fundamentals of Java.

Java Servlets

In order to make use of JavaServer Pages to supply dynamic page content, it is necessary first to understand how Java servlets work; see http://java.sun.com/products/servlet for an overview of Java servlet technology. Java servlets are the server-side equivalent of Java client-side applets (see Appendix B, "Client-Side Programming with Java Applets," for a discussion of writing Java applets). Like JavaScript and applets, servlets can be run only within a *servlet container*, and are called from a Web browser. There is no main method as in Java console applications. A user calls the servlet from the client Web browser, the servlet executes in the servlet engine, and then it sends results back to the browser (in the form of HTML code). Consequently, Java servlets require a servlet engine, or container such as Apache Jakarta Tomcat, IBM's WebSphere or BEA WebLogic; see http://servlets.com/engines/ for a detailed list of commercial and open-source servlet engines).

The function of the servlet engine is to load the servlet .class file into the Java Virtual Machine (JVM) running on the server. The engine can then run the servlet. (The .class file is not reloaded into the JVM again after the first time; most servlet engines now include options to reload .class files automatically when they are updated, so that it is usually necessary to restart the server when the .class file changes.

The most widely available servlet engine is Tomcat from the Apache Foundation Jakarta Project (see http://jakarta.apache.org/tomcat). Since it is both free and widely used, the examples presented in this article all use the Tomcat engine with the Apache Web server.

It is to be hoped that users at sites with other server software will be able to generalize from the principles and requirements illustrated here.

The Tomcat servlet engine is available for a wide variety of platforms; commonly, it can be installed on either Windows or UNIX systems.[1] For the examples that follow, the instructions reference a server running a version of the Linux operating system, but they apply equally to a Windows host as well. In order to configure Apache or IIS to use the Tomcat servlet container, it is first necessary to install the Tomcat redirector plug-in.[2] However, Tomcat is itself an HTTP server, so it can work without Apache or IIS.

A Hill of Beans

One additional concept needs to be introduced in order for you to understand how servlets and JSP can be customized. The JavaBeans architecture provides the developer with a set of self-contained components that interoperate according to a standard set of rules. These are the cross-platform equivalent of the Microsoft Component Object Model (COM). If you have created forms in Microsoft Access or Visual Basic, you are familiar with the process of selecting controls from a toolbox and dropping them onto the form. Of course this is just a small part of COM, which provides users of Microsoft software the ability to build and deploy all sort of reusable components.

For developers who wish to create non-proprietary solutions, the answer is to use JavaBeans. Do not confuse the JavaBeans architecture with *Enterprise JavaBeans*, which are a way to distribute components across all the platforms in an organization (and hence equivalent to Microsoft's DCOM). JavaBeans are just reusable software components. They are frequently used in visual programming environments (like webAF) but it is not necessary for a Bean to have a visual component.

The following example shows what is perhaps the simplest possible Bean. It has only one data item, an integer field called num, and two methods, one to set the value of num and the other to get it.

Example 8.1 The Basic JavaBean

```
// A simple Bean
import java.io.Serializable;

public class TestBean
  implements Serializable
{
  private int num;

  public void setValue(int n)
  {
      num = n;
  }
```

---

[1] The Tomcat binaries are frequently updated as new features are added. Note that while the current version is Tomcat 5.5, Tomcat 4.1.18 is the only one recommended for use with SAS AppDev Studio 3.1 (see the next chapter for more details).

[2] See Gal Shachor, "Tomcat IIS HowTo," The Apache Software Foundation, 1999-2001. http://jakarta.apache.org/tomcat/tomcat-3.3-doc/tomcat-iis-howto.html.

```
public int getValue()
{
    return num;
}
}
```

A good Bean has to follow the rules: the two most important are that Beans must be *serializable* and that they must conform to a specific set of *design patterns* when naming features (like the `getValue()` and `setValue()` methods in the example). This enables them to interoperate with other components. The term "serializable" means that you can store a Bean in a file on the computer or send it over a network to another host, an obvious necessity for components that are used to create distributed systems. The second requirement ensures that Beans will all work together in harmony, without the developer needing to know (or care) what is happening inside the Bean code.

For a more detailed overview of the JavaBean architecture, see http://java.sun.com/docs/books/tutorial/javabeans/whatis/beanDefinition.html. Sun has also provided the Beans Development Kit (BDK) free to developers; you can download it from their Web site at http://java.sun.com/products/javabeans/index.jsp. The BDK enables you to write your own JavaBeans (assuming you can program in Java).

Fortunately for developers, you do not need to be able to create your own Beans in order to be able to use the full power of this new technology. In Chapter 9, we shall see how to use webAF 3.1 software and the reusable components that SAS has provided to create powerful, modular, and reliable Web servlets.

Servlet Example

The canonical first program in any software tutorial is always "Hello World;" here is a servlet example:

Example 8.2 Hello World Servlet

```
import java.io.*;
import javax.servlet.*;
import javax.servlet.http.*;

public class HelloWorld extends HttpServlet
{
    public void doGet (HttpServletRequest request,
        HttpServletResponse response)
        throws ServletException, IOException
    {
        PrintWriter out;
        String title = "Simple Servlet Output";

        response.setContentType("text/html");
        out = response.getWriter();

        out.println("<HTML><HEAD><TITLE>");
        out.println(title);
        out.println("</TITLE></HEAD><BODY>");
        out.println("<H1>" + title + "</H1>");
        out.println("<P>Hello World!");
```

```
        out.println("</BODY></HTML>");
        out.close();
    }

    public void doPost (HttpServletRequest request,
        HttpServletResponse response)
        throws ServletException, IOException
    {
        doGet(request, response);
    }
}
```

This class illustrates the most important features of servlet programming. The servlet class is derived from the parent class `javax.servlet.http.HttpServlet.` The two overridden methods are:

- `doGet()` – called either by going directly to the page or by an HTML `get` request
- `doPost()` – called by a form using a `post` request

In this case, both method calls result in the same code being executed. The `HttpServletRequest` parameter object is bound to the page that called the servlet, while the `HttpServletResponse` parameter is the generated page. The servlet writes to a `PrintWriter` object that is bound to the response object by the call to `getWriter()`; the `out.println` statements then generate the HTML source.

The Tomcat engine must be running as a server process; this is the responsibility of the system administrator, who can configure the server to start Tomcat automatically at start-up. Assuming Tomcat is started, the HTML output of this program should appear be as shown in Display 8.1:

Display 8.1 Hello World Servlet Output

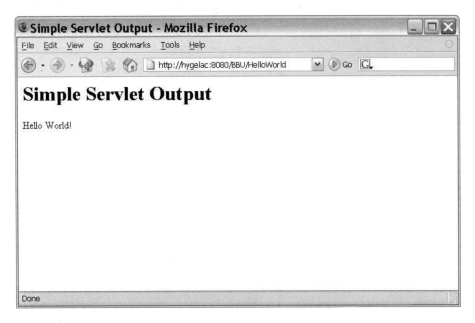

The default Tomcat installation creates additional samples in http://<*server-name*>:8080/servlets-examples/, where <*server-name*> is the name of your Tomcat server. There is even a Hello World servlet, similar to the preceding example, along with samples of how to use URL parameters and cookies to supply state information to the server.

Deploying Servlets

Note that the URL for the preceding Web page references an instance of Tomcat running on port 8080. One important but frequently overlooked issue in explaining Web application deployment is the directory structure of the Web server. While the Apache Web server usually listens on port 80 by default, Tomcat uses a different TCP/IP port. They can also be set up to work together, instead of interchangeably; that is, Apache can be set up so that it forwards certain requests to Tomcat. This requires the addition of the mod_jk dynamic module to Apache; see the Apache Foundation Web site for directions on how to accomplish this on various operating system platforms.

Tomcat uses the default directory **$CATALINA_HOME/webapps**, where **$CATALINA_HOME** is the folder where Tomcat is installed. The simplest way to deploy servlet content is first to create a subdirectory under webapps. In this example it is called **$CATALINA_HOME/webapps/BBU**. The servlet code is located under this directory; the following diagram illustrates the required structure (the directory names are case sensitive).

Display 8.2 Tomcat Servlet Deployment

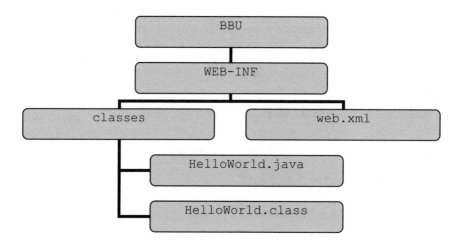

In order to compile the Java code, be sure to include the **servlet-api.jar** file in the classpath; for example, starting from the **$CATALINA_HOME** folder:

```
javac -classpath common/lib/servlet-api.jar
    webapps/BBU/WEB-INF/classes/HelloWorld.java
```

The other file listed, **web.xml**, is used to specify the application to the servlet container. The following example shows a minimal configuration; for more information on setting up and configuring Tomcat using XML descriptor files, see http://jakarta.apache.org/tomcat/tomcat-4.1-doc/

Example 8.3 Servlet Configuration File web.xml

```
<?xml version="1.0" encoding="ISO-8859-1"?>
<!DOCTYPE web-app PUBLIC
    "-//Sun Microsystems, Inc.//DTD Web Application 2.3//EN"
    "http://java.sun.com/dtd/web-app_2_3.dtd">
<web-app>
  <display-name>
       BBU Examples
  </display-name>
      <description>
              Hello World servlet example
  </description>
  <servlet>
              <servlet-name>HelloWorld</servlet-name>
              <servlet-class>HelloWorld</servlet-class>
  </servlet>
  <servlet-mapping>
              <servlet-name>HelloWorld</servlet-name>
              <url-pattern>/HelloWorld</url-pattern>
      </servlet-mapping>
</web-app>
```

The **web.xml** file is used to configure the servlet container to recognize the application; the preceding example contains four elements (from among the extensive set of possible options):

- `display - name` the name of the Web application
- `description` - a description of the Web application
- `servlet` - the name of the Java .class file
- `servlet-mapping` - the URL mapping

After placing the file in the **WEB-INF** folder, stop and restart the Tomcat server. The application will be loaded and the URL should work.

Creating a JavaServer Page

To create a JSP, first build an XHTML page file named with a .jsp extension instead of .html. Now add three kinds of JSP elements:

- Scripting elements – Java code that becomes part of the servlet
- Directives – code that controls the overall structure of the servlet
- Actions – code that specifies existing components that should be used

Deploy the file to the correct location on the server and open in your favorite browser. That's it! Of course, in practice it's a little more complicated than that.

JavaServer Pages combine static HTML, known as *template text*, with dynamically generated content. Within the HTML, Java code is enclosed in special tags. The JSP container compiles the Java code (the first time only) and creates a .class file in the **work** directory on the server. Thus running JSP is a three-step process: the servlet engine generates a **java** source code file that is then compiled and executed. (In production environments, the pages are usually precompiled after they are loaded on the server. In this way, the first user of the page is not confronted with a lengthy delay while first the source and then the byte code is generated.)

Scripting Elements

Three kinds of scripting elements are used in JSP.

Expressions

Expressions must evaluate to type `java.lang.String` since they are inserted directly into `java.lang.println` statements. The initial symbol must be `<%=` while the closing symbol for all scripting elements is `%>`.

JSP code: `<%= java.util.Date() %>`

Generated servlet code: `out.println(java.util.Date());`

Scriptlets

Scriptlets are in-line Java code. The initial symbol must be `<%` while the final symbol is `%>`. The code is inserted into the generated servlet's `_jspService` method to determine dynamically what text, graphics, links, etc. are displayed. As in CGI, parameters are passed automatically via the URL and accessed in JSP through the `request` object (see the following code). Scriptlets are useful for making parts of JSP file conditional. For example:

```
<% if (request.getParameter("txtBox") == null)
   {
       out.println("You need to enter a value!");
   }
   else
   {
       out.println(txt);
   }
%>
```

Declarations

The initial symbol must be `<%!`; the final symbol is `%>`. Declarations are inserted into the body of the servlet outside of the `_jspService` method. Typically they do not generate any output and are used in conjunction with expressions and scriptlets.

```
<%! private int accessCount = 0; %>
Accesses to page since server reboot: <%= ++accessCount %>
```

Predefined Variables

The JSP container provides several predefined objects for use in scriptlets, including the following:

- `request` – the `HttpservletRequest` object
- `response` – the `HttpservletResponse` object
- `session` – the `HttpSession` associated with the request
- `out` – the `PrintWriter` used to send output to the client

Directives

JSP directives control the overall structure of the servlet. There are three kinds of directives: `page, include` and `taglib`. The required initial symbol is `<%@`; the final symbol is `%>`.

The `page` directive is used to specify various attributes that are included in the code to describe the output. Some of the available `page` attributes and their possible values are shown in the following table:

Table 8.1 Page Directive Attributes

Page Attribute	Value
`import`	`package.class`
`contentType`	`MIME-Type`
`isThreadSafe`	`true|false`
`session`	`true|false`
`buffer`	`sizekb|none`
`autoflush`	`true|false`
`extends`	`package.class`
`info`	`message`
`errorPage`	`URL`
`isErrorPage`	`true|false`
`language`	`java`

The `include` directive in JSP is used to insert an external file into the page at the specified point. Note that when using the `include` directive, the file is inserted when the page is initially compiled, not at execution time. The `include` action (see the following section) is used to insert content when the page is executed. The following is an example of the `include` directive:

```
<%@ include file="menu.jsp" %>
```

The `taglib` directive references a custom tag library; see "Using Custom Tags" later in this chapter for an explanation of how this directive is employed.

Actions

JSP actions may employ XML syntax to specify existing components that should be used. An action must have a start tag which must use a prefix. If the body is empty, as it frequently is, the action can use the empty tag syntax. For the following standard actions the prefix must be `jsp`. For custom tags it can be any name defined by the JSP `taglib` directive. The following list illustrates the most common actions:

- `<jsp:include page="MyPage.jsp" flush="true" />`
- `<jsp:forward page="/utils/errorReporter.jsp" />`
- `<jsp:useBean id="myBean" class="MyBeanClass"/>`
- `<jsp:setProperty name="myName" property="someProperty"/>`
- `<jsp:getProperty name="itemBean" property="numItems" />`

As noted above, the `include` action has an effect similar to the `include` directive; the difference is that the action occurs when the page is loaded and not when the code is generated. This distinction allows for more dynamic rendering.

The `forward` action redirects the browser to a new page, either JSP or HTML.

Properties

JSP uses `getProperty` and `setProperty` methods to manage properties. There are two ways to load a property with a value, using either XML syntax or object dot notation in a scriptlet or expression:

- `<jsp:setProperty name="name" property="property"  value="value" />`
- `<%= name.setProperty("value"); %>`

There are also two ways to get the value of a property:

- `<jsp:getProperty name="name" property="propName" />`
- `<%= name.getPropName(); %>`

Parameters

There are two ways to load a property with a parameter value:

- `<jsp:setProperty name="entry" property="itemID"`
  `       value='<%= request.getParameter("itemID") %>' />`
- `<jsp:setProperty name="entry" property="itemID"`
  `       param="itemID" />`

The first approach uses the `getParameter` method of the `HttpServletRequest` object to get the values of parameters passed to the page by the HTTP `get` or `post` methods. The second way takes advantage of the `param` attribute of the JSP `setProperty` tag as a shortcut to the value of the named parameter. Types are automatically converted, since all parameters are passed as text.

Scope

JSP elements can also have an optional `scope` attribute. This defines the scope of the Bean and allows a single class to be used throughout the Web application or the HTTP session. The four possible values for scope are:

- `page` – the default
- `application` – any servlet in the same Web application
- `session` – the current HTTP session
- `request` – the servlet `request` object (usually equivalent to `page`)

A Simple JSP Example

Here is another version of the Hello World servlet, rewritten as a JavaServer Page:

Example 8.4 Simple JSP Example

```
<?xml version="1.0"?>
<!DOCTYPE html PUBLIC "-//W3C//DTD XHTML 1.0 Transitional//EN"
   "http://www.w3.org/TR/xhtml1/DTD/xhtml1-transitional.dtd">
<html xmlns="http://www.w3.org/1999/xhtml">
<head>
   <title>JSP Examples</title>
   <style type="text/css">
       h1    { color: blue; }
       body { font-family: helvetica, arial; }
   </style>
</head>
<body>
   <h1>Simple JavaServer Page</h1>
   <% out.print("Hello World!"); %>
   The time now is <%= new java.util.Date() %>
</body>
</html>
```

Note that although this example uses XHTML, this is not necessary for JSP; conventional HTML 4.0 works just as well. The example also includes an inline style sheet. As discussed below, most of the tags for formatting text (specifically the `<font>` tag) were deprecated in HTML 4.01 and are not supported in XHTML 1.0 Strict DTD. The standard for HTML formatting has become CSS.

The entire page is template text except for the two lines shown in bold, which illustrate two different methods for including Java code in JSP. The statement

```
<% out.print("Hello World!"); %>
```

is a *scriptlet*—that is, embedded Java code. The next line

```
The time now is <%= new java.util.Date() %>
```

is just HTML with the addition of a JSP *expression*.

The output of this program is shown in Display 8.3:

Display 8.3 Simple JSP Example Output

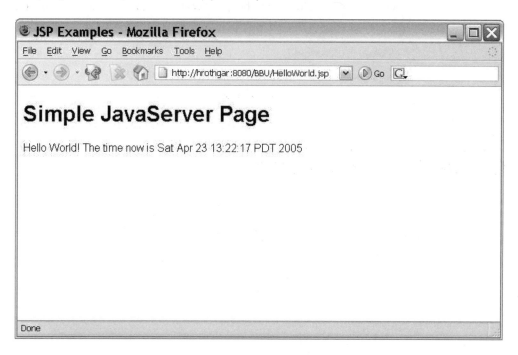

As noted in the section on servlets, the default directory for serving pages in Tomcat is **$TOMCAT_HOME/webapps**. Tomcat uses the file **$TOMCAT_HOME/conf/server.xml** to configure the behavior of the Tomcat JSP container. The **server.xml** file can be modified to change the default directory (see http://tomcat.apache.org/tomcat-4.0-doc/config/ for an explanation of how the Tomcat configuration files work). As a result, the actual name of the page in the example above is not the same as the URL shown in the output:

```
URL:        <server_name>:8080/BBU/HelloWorld.jsp
Pathname:   $TOMCAT_HOME/webapps/BBU/HelloWorld.jsp
```

This has been determined to lead to considerable confusion when first deploying JSP applications!

Note that selecting View Source on this page will not display the JSP code, but instead the HTML generated by the servlet. The only way to view the generated servlet code is by looking at the following file (the exact location of the generated code will vary by server, and perhaps by release of server):

```
$CATALINA_HOME/work/Catalina/localhost/BBU/
   org/apache/jsp/HelloWorld_jsp.java
```

The servlet container manages the process of converting the JavaServer Page to a servlet, compiling it with the appropriate class libraries, and executing the resulting code using the Java Virtual Machine on the local host. Instead of manually writing Java code, the developer need only insert the correct directives and actions.

JSP Custom Tag Libraries

In order to create fully reusable distributed components, Java now provides support for custom JSP *tag libraries*; see http://java.sun.com/products/jsp/jstl/reference/docs/. The advantage of using custom JSP tags is that Web development may be separated into two components: *presentation* and *logic*. The work of creating the HTML template text can be accomplished by multimedia specialists, while the components that provide the underlying functionality can be developed independently by Java programmers. By supplying the HTML coders with custom JSP tags, the entire process can be streamlined and made somewhat more robust.

The JSP specification provides for a set of standard actions to access objects and their properties. Actions are written using XML syntax (see the preceding discussion). In addition, however, the developer can create custom actions that have a wide range of functionality. In order to use custom tags, three components are necessary:

- the *tag handler* class, which specifies the tag's behavior
- the *tag library descriptor* (TLD) file, which maps XML element names
- one or more pages that use the tag library

Tag Handlers

Tag handlers are a type of Java event handler that is similar to the SAX-2 XML Parser interface. They are specialized Java classes that implement the `javax.servlet.jsp.tagext.tag` interface or one of its sub-interfaces. The tag handler is invoked by the JSP container to implement the action specified by a custom tag. Some of the methods available in the `tag` API are:

- `doStartTag()` – processes the start tag for this instance
- `doEndTag()` – processes the end tag for this instance
- `getParent()` – gets the parent (closest enclosing tag handler) for this tag handler
- `setParent()` – sets the parent (closest enclosing tag handler) of this tag handler

For a simple tag without attributes or a body, the developer simply overrides the `doStartTag()` with the code to be executed. When the JSP container encounters the tag, the generated servlet code will come from the referenced tag method. Examples of each of the three required files are given below.

The following Java program can be used to print out "Hello World" and the current date and time.

Example 8.5 Custom Tag Handler

```
import java.util.Date;import javax.servlet.jsp.*;
import javax.servlet.jsp.tagext.*;

public class SimpleTag
  extends TagSupport
  {
  public int doStartTag() throws JspException
  {
```

```
    try
    {
        pageContext.getOut().print
        ("Hello World." + "The time now is " +
         new Date());
    }
    catch (Exception ex)
    {
    throw new JspTagException
        ("SimpleTag: " + ex.getMessage());
    }

    return SKIP_BODY;
    }

    public int doEndTag()
    {
    return EVAL_PAGE;
    }
}
```

The `doStartTag()` method is triggered when the start tag is encountered, sending the message to the client. The two possible return codes for `doEndTag()` are `EVAL_PAGE`, indicating that page evaluation should continue, and `SKIP_PAGE`, which causes the code on the remainder of the page to be bypassed.

Tag Library Descriptors

A *tag library descriptor* (TLD) file has the extension `.tld`. A TLD is just an XML document that describes a tag library. The syntax is quite complex, however; see the online documentation for the XML Document Type Definition (DTD). There were significant changes between Version 1.1 and Version 1.2 of the DTD; be sure to use the correct specification. See http://java.sun.com/dtd/ for a list of the available J2EE downloads.

The function of the TLD is to define the custom tags in the JSP. One possible TLD for the preceding example is shown in Example 8.6:

Example 8.6 Tag Library Description

```
<?xml version="1.0"?>
<!DOCTYPE taglib PUBLIC
    "-//Sun Microsystems, Inc.//DTD JSP Tag Library 1.2//EN"
    "http://java.sun.com/dtd/web-jsptaglibrary_1_2.dtd" >
<taglib>
        <tlib-version>1.0</tlib-version>
        <jsp-version>1.2</jsp-version>
        <short-name>Sample tag library</short-name>
        <description>Library for simple tag example
        </description>
        <tag>
                <name>SayHello</name>
                <tag-class> SimpleTag</tag-class>
                <tei-class>EMPTY</tei-class>
                <description>Hello world example</description>
        </tag>
</taglib>
```

This XML file references the JSP 1.2 Tag Library Descriptor (TLD) at http://java.sun.com/dtd/web-jsptaglibrary_1_2.dtd. Note that although this looks like a reference to a Web page, in fact the URI is used here as a label for the set of definitions that define the allowable tags. This Document Type Definition must be included in the TLD or it won't work.

The element names are as specified in the DTD; the root element is `<taglib>`, and one or more `<tag>` elements may be included. This TLD includes only one tag, called `SayHello`.

JSP with Custom Tags

A Web page that uses custom tags must include a `taglib` directive referencing this file before any custom tag is used. For example:

```
<%@ taglib URI="sample.tld" prefix="test" %>
```

The URI attribute may point to a file in the same directory as the JSP, in another directory, or on any host accessible to the server. The URI will often be a reference that is defined in the `web.xml` file. As the Sun documentation points out, the tag prefix is

> …used to distinguish a custom action, e.g. `<myPrefix:myTag>`. Prefixes starting with `jsp:`, `jspx:`, `java:`, `javax:`, `servlet:`, `sun:`, and `sunw:` are reserved. A prefix must follow the naming convention specified in the XML namespaces specification and empty prefixes are illegal in this version of the specification. (http://java.sun.com)

Just as the `jsp:useBean` element uses the `id` attribute to specify a prefix for Bean properties and methods, so `taglib` uses `prefix` to identify the XML namespace for custom tags. JSP custom tags are then written using this prefix:

```
<test:tag>body</test:tag>
```

For simple tags with no body, the prefix is just `<test:tag />`.

Tags can and frequently do have attributes, and attribute values can be set by run-time expressions.

```
<logic:iterate collection="<%=bookDB.getBooks()%>"
      id="book" type="database.BookDetails">
```

A page using this TLD and handler might look like the following:

Example 8.7 JSP Using Custom Tags

```
<?xml version="1.0"?>
<!DOCTYPE html PUBLIC "-//W3C//DTD XHTML 1.0 Transitional//EN"
  "http://www.w3.org/TR/xhtml1/DTD/xhtml1-transitional.dtd">
<html xmlns="http://www.w3.org/1999/xhtml">
<%@ taglib URL="simple.tld" prefix="test" %>
<head>
  <title>JSP Examples - Custom Tag Library</title>
  <style type="text/css">
      h1   { color: blue; }
      body { font-family: helvetica, arial; }
  </style>
</head>
```

```
<body>
  <h1>JavaServer Page Custom Tag Example</h2>
  <test:SayHello />
</body>
</html>
```

Deploying Tag Libraries

Custom tags are generally collected into packages; in the previous example, the class `SimpleTag` is part of the `examples` package. Consequently, in `simple.tld`, the tag library descriptor file, the `SimpleTag` class is referenced as `examples.SimpleTag`.

It is important to understand where each of these files is located on the server.

- The JSP file is located under the Tomcat server root directory as follows:

  Pathname: **$TOMCAT_HOME/webapps/examples/jsp/example3.jsp**
  URL: **<server-name>:8080/examples/jsp/example3.jsp**

- The TLD file can be anywhere the server can find it; in this case it is stored in the same directory as the JSP, so the URL in the `taglib` directive simply refers to `simple.tld`. It may be better to use the URI in your taglib, which refers to a taglib entry in your **web.xml** file. Using this approach, you will not have to hardcode paths in each JavaServer Page.

- The Java .class file must be in the servlet **WEB-INF/classes** subdirectory. Since it is in the package `examples`, the path to this file is as follows:

  **$TOMCAT_HOME/webapps/examples/WEB-INF/classes/
  examples/SimpleTag.class**

Additional custom tags can be referenced in the TLD file, and the associated class files must also appear in this directory.

Database Access Using JDBC and JavaServer Pages

One of the most common applications for JSP is reading and writing to databases. This is usually accomplished via *Java Database Connectivity* (JDBC) the Java API for connecting to different kinds of data sources. For detailed information about JDBC, see Sun's documentation at http://java.sun.com/products/jdbc/. Although this topic is not strictly speaking part of the JSP specification, it is important to understand how Java database code can be implemented using JSP and particularly custom tags.

JDBC Overview

The JDBC API contains two packages:

- **java.sql** – the basic JDBC package
- **javax.sql** – a package that adds some additional server-side capabilities

This review will only consider the classes and methods of the first package, **java.sql**, since this illustrates most of the concepts that are required for using JDBC with JSP. For more information about the **javax.sql** package, see the API documentation available at the Sun Web site.

The first task in implementing a JDBC application is to choose a *driver*. These are available from various vendors, and are specific to a particular database management system (see http://industry.java.sun.com/products/jdbc/drivers). These drivers must be downloaded, installed on the server, and added to the Java classpath.

As noted in Chapter 4, "Remote Access to SAS," there are three different kinds of SAS servers. For this example, we will use the JDBC driver for SAS/SHARE*NET; additional SAS drivers are available for SAS/CONNECT software and the Integration Object Model (IOM). Since Java is designed for developing cross-platform applications, there are also other drivers for database systems such as Oracle, DB2, Sybase, and even Microsoft Access (the JDBC-ODBC bridge). Modifying a Java program to access a different database is usually just a matter of changing a few lines of code to select a different driver.

Downloading and Installing a JDBC Driver

Depending on the products licensed, the SAS drivers are generally included on the installation CDs. If not, they are also available from the SAS support site at http://support.sas.com/apps/demosdownloads/setupintro.jsp. Select "SAS Drivers for JDBC" and you should get a link to the download page. The drivers for Windows are available as **jdbcdrivers_win.tar**; Linux drivers are in **jdbcdrivers_lnx.tar**, and so forth. You will need to register with SAS to download one of these files, but the process is free and only takes a moment.

The installation files can be unpacked using WinZip or the appropriate decompression application for your platform. Three package files are created: **sas.core.jar**, **sas.svc.connection.jar**, and **sas.intrnet.javatools.jar**. These must be added to your Java classpath for the program shown in Example 8.8 to work.

Obtaining a Connection

In order to implement JDBC in a program, the first step after binding to a driver is to create a connection object. The following example illustrates how to open a connection to SAS using the JDBC driver for SAS/SHARE software.

Example 8.8 JDBC Example

```
import java.sql.*;

/* connection test to a remote SAS/SHARE server using JDBC */
public class JDBCTest
{
  public static void main(String[] args)
  {
      Connection con = null;

      try // open connection to database
      {
            Class.forName(
                "com.sas.net.sharenet.ShareNetDriver");
```

```
                    con = DriverManager.getConnection(
                            "jdbc:sharenet://hunding:8551?
                            user=sasdemo&password=sasuser");

                    // print connection information
                    DatabaseMetaData dma = con.getMetaData();
                    System.out.println(
                            "Connected to " + dma.getURL());
                    System.out.println(
                            "Driver " + dma.getDriverName());
                    System.out.println(
                            "Version " + dma.getDriverVersion());
            }
            catch(Exception e) { e.printStackTrace(); }
            finally // make sure connection gets closed properly
            {
                    try
                    {
                            if (null != con) con.close();
                    }
                    catch (SQLException e) {}
            }
        }
    }
```

For this simple example, the program just opens a `java.sql.DatabaseMetaData` object for the connection and prints some diagnostics.

Running this program should display the following message (in the Windows environment):

```
Connected to jdbc:sharenet://hunding:8551
Driver SAS/SHARE driver for JDBC, Version 9.1 Production
Version 9.1
Press any key to continue . . .
```

The SAS/SHARE JDBC driver is in **sas.intrnet.javatools.jar**; as previously noted, this package file must be in your Java classpath. More information about using the SAS/SHARE JDBC driver is available at http://support.sas.com/rnd/web/intrnet/java/jdbc/.

It is important to make sure the connection is closed explicitly, since otherwise "zombie" processes may be created on the database server.

You can also use the Integrated Object Model (IOM) as a JDBC data connection. Setting up an IOM object spawner is described in detail in the chapters that follow. For now, note that only two changes need to be made in the previous program: the driver class name for IOM is `com.sas.rio.MVADriver`, and the connection string is `jdbc:sasiom://<host-name>:8591,"<username>","<password>"`. The IOM driver is in **sas.svc.connection.jar**; this package file must be in your Java classpath.

The procedures for connecting using SAS/CONNECT software are somewhat more complex; see http://support.sas.com/rnd/web/intrnet/java/ for information about how to set up a JDBC connection to a SAS/CONNECT server.

Using SQL

In order to read from or write to the database, JDBC uses SQL syntax. There are several types of objects available, the most important of which are the following:

- `Statement` – executes a simple statement with no parameters
- `ResultSet` – holds the results of a SELECT query
- `PreparedStatement` – optimizes repeated queries
- `CallableStatement` – calls database stored procedures

A `Statement` object is used to send SQL statements to the database: `PreparedStatement` and `CallableStatement` are actually both kinds of *statements*.

A `Select` statement creates a `ResultSet` object, as shown in the following example:

```
Statement stmt = con.createStatement();
ResultSet rs = stmt.executeQuery
   ("SELECT a, b, c FROM Table2");
```

There are three different methods for executing SQL statements:

- `executeQuery` – for statements that produce a single `ResultSet`
- `executeUpdate` – for INSERT, UPDATE, or DELETE statements
- `execute` – for complex queries

A `ResultSet` object is the result of executing an SQL SELECT query. Database fields are returned by various `getXXX` methods, as shown below:

```
ResultSet rs = stmt.executeQuery
   ("SELECT a, b, c FROM Table1");
while (rs.next())
{
   int i = rs.getInt("a");
   String s = rs.getString("b");
   float f = rs.getFloat("c");
   System.out.println("ROW = " + i + " " + s + " " + f);
}
```

If the `ResultSet` is empty, the return value of the Boolean method `getNext()` is false; otherwise the effect of the method is to move the result set cursor to the next row. When the end of the set is reached, the value of `getNext()` is false.

Using SQL with JSP Custom Tags

There are a number of freely available tag libraries, including several specifically for database access. As the Sun documentation points out:

> The JavaServer Pages Standard Tag Library (JSTL) encapsulates, as simple tags, core functionality common to many JSP applications.... This standardization lets you learn a single tag and use it on multiple JSP containers.
> (http://java.sun.com/products/jsp/jstl/)

The JSTL tags are designed for prototyping and not for production use. For industrial-strength applications, Sun recommends encapsulating the functionality in JavaBeans components. (This is in fact precisely what SAS has done in SAS AppDev Studio.) Nonetheless, it is instructive to look at an example of some of the SQL tags available in the standard library.

The Web page shown in Display 8.4 can be constructed using JSTL with a minimum of Java code. The example displays the familiar SASHELP.CLASS data set as an HTML table. Note that this page is not editable, but it is updated every time the page is opened.

Display 8.4 JSTL Example

The code for this example is shown in Example 8.9. The first two lines reference the two required tag libraries from the Apache Jakarta project: the *core* library tags are referenced with the prefix "c" while the *sql* library uses the prefix "sql".

Example 8.9 Using JDBC with JSTL Custom Tags

```
<%@ taglib prefix="c" uri="http://java.sun.com/jstl/core" %>
<%@ taglib prefix="sql" uri="http://java.sun.com/jstl/sql" %>

<html>
<head>
  <title>JSP Examples - JDBC</title>
  <style type="text/css">
      h1    { color: maroon; text-align: center; }
      h2    { color: blue; text-align: center;}
      body  { font-family: helvetica, arial; }
```

```
        td.c    { text-align: center; }
        td.r    { text-align: right; }
        th      { background-color: yellow; width: 10%; }
    </style>
</head>

<body>
    <h1>SASHELP.CLASS</h1>
    <h2>SAS Share JDBC Driver</h2>

    <table border="1" align="center" id="class dataset">

        <%-- open a database connection --%>
        <sql:setDataSource
            var="datasource"
            driver="com.sas.net.sharenet.ShareNetDriver"
            url="jdbc:sharenet://hunding:8551?
                user=sasdemo&password=sasuser"
        />

        <%-- execute the database query --%>
        <sql:query
            var="students"
            dataSource="${datasource}" >
            select * from sashelp.class</sql:query>

        <%-- Print the column names for the header of the table --%>
        <tr>
            <th>Name</th>
            <th>Age</th>
            <th>Sex</th>
            <th>Height</th>
            <th>Weight</th>
        </tr>

        <%-- loop through the rows of the query --%>
        <c:forEach var="row" items="${students.rows}">
          <tr>
                <td><c:out value="${row.name}"/></td>
                <td class="r"><c:out value="${row.age}"/></td>
                <td class="c"><c:out value="${row.sex}"/></td>
                <td class="r"><c:out value="${row.height}"/></td>
                <td class="r"><c:out value="${row.weight}"/></td>
          </tr>
        </c:forEach>
    </table>
</body>

</html>
```

The example program uses four JSTL tags to encapsulate the JDBC database actions described previously. The rest of the program is just standard HTML. A little CSS is used to format the table. The four tags are as follows:

- `sql:setDataSource` – creates a connection to the JDBC data source. As the Jakarta documentation warns, you should not use this method for production Web sites. The details of using JINDI to manage connections are beyond the scope of this discussion.
- `sql:query` – passes the SQL select statement through to the SAS server. In this case, 20 records are selected.
- `c:forEach` – loops over the result set, creating one table row for each observation.
- `c:out` – prints the variable values.

Deploying a JavaServer Page Containing JSTL

In order to deploy this example, copy the JavaServer Page file **JDBCTest.jsp** to the desired directory in the **webapps** folder on the server. As shown in Display 8.2, for this example we created a folder called **BBU**. Now make sure that the necessary JAR files are all loaded into the **lib** subdirectory of **WEB-INF** under this. For this example, the absolute pathname of the Web application folder is **${CATALINA_HOME}/webapps/BBU/WEB-INF/lib**.

The following five Java package files are required:

- **sas.core.jar**
- **sas.internet.javatools.jar**
- **sas.svc.connection.jar**
- **jstl.jar**
- **standard.jar**

The three SAS files in this list can be copied from the **SASDriversforJDBCandCONNECT** directory created at setup; see the previous section "Downloading and Installing a JDBC Driver" for the details about obtaining these files. The last two files can be obtained from the Apache Jakarta project at http://jakarta.apache.org/site/downloads/downloads_taglibs.html. Download and unzip the packages into a convenient directory. As noted above, you will need to make copies of these two files in the **WEB-INF/lib** subdirectory in your **webapps** application folder in order to get your JSP to work.

One other aspect of using JDBC with SAS needs to be pointed out: the question of setting the Java permissions for the servlet engine. You need to edit the file **${CATALINA_HOME}/conf/catalina.policy** so Tomcat can access the SAS server. The simplest thing to do is just to give the Web application unlimited privileges. To test the page in Example 8.9, add the following to the end of your policy file:

```
grant codeBase "file:${catalina.home}/webapps/BBU/-" {
    permission java.security.AllPermission;
  };
```

In a production environment, you will need to work with your Web administrator to set the correct permissions.

With JSTL, the Web designer does not need to know the details of how the tags work, but only the correct syntax for their attributes. In Chapter 9, you will see how webAF software can be used to construct JSP pages using Java *InformationBeans* that supply built-in access to SAS data sets. First, though, it might be worthwhile to look at one more use of JSTL.

Building a Data Entry Form with HTML and JSTL

The following example is useful for a simple online survey, a registration system, or any other application that uses an HTML form to capture data from users. The most common way to set up this application is in two parts: first, an HTML page is used to collect the data, and second, a JavaServer Page can be used to update the database. Example 8.10 shows a very vanilla data entry form, coded in HTML.

Example 8.10 HTML Data Entry Form

```html
<html>
<head>
  <title>JSP Examples: JSTL Demo</title>
</head>
<body style="font-family: verdana, tahoma, ariel;">
<div style="width: 50%; margin-left: auto; margin-right: auto;">
  <h1 style="color: #0000FF;">Demo Entry Form</h1>
  <form method="get" action="update.jsp">
      <table border="0" id="input form">
          <tr>
                  <td style="font-weight: bold;">Name</td>
                  <td><input type="text" name="name" /></td>
          </tr>
          <tr>
                  <td style="font-weight: bold; vertical-
                  align: top">Gender</td>
                  <td>
                          <input type="radio" name="sex"
                          value="M" >Male<br />
                          </input>
                          <input type="radio" name="sex"
                          value="F">Female
                          </input>
                  </td>
          </tr>
          <tr>
                  <td style="font-weight: bold;">Age</td>
                  <td><input type="text" name="age" /></td>
          </tr>
          <tr>
                  <td style="font-weight: bold;">Height</td>
                  <td><input type="text" name="height" /></td>
          </tr>
          <tr>
                  <td style="font-weight: bold;">Weight</td>
                  <td><input type="text" name="weight" /></td>
          </tr>
      </table>
```

```
        <p>
                <input type="submit" value="Submit"/>
                <input type="reset" value="Reset" />
        </p>
    </form>
</div>
<body>
</html>
```

When displayed in a Web browser, the form looks something like Display 8.5.

Display 8.5 Demo Entry Form in HTML

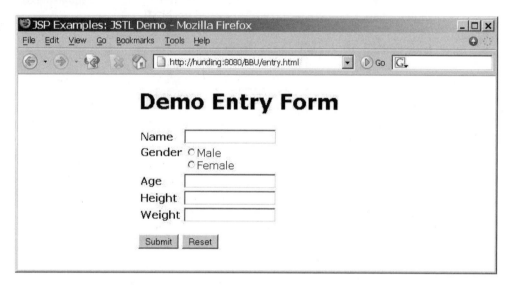

Pressing the **Submit** button loads the following JavaServer Page:

Example 8.11 Updating a Database with JSTL and JDBC

```
<%@ taglib prefix="sql" uri="http://java.sun.com/jstl/sql" %>
<html>
<head>
  <title>JSP Examples: JSTL Demo</title>
</head>

<%-- update data --%>
<%
  /* SQL code to append parameters to CLASS table */
    String strSQL="INSERT INTO sashelp.class SET " +
        "name=\""+request.getParameter("name")+"\","+
        "sex=\""+request.getParameter("sex")+"\","+
        "age="+request.getParameter("age")+","+
        "height="+request.getParameter("height")+","+
        "weight="+request.getParameter("weight");
%>

<sql:setDataSource
        var="class"
        driver="com.sas.rio.MVADriver"
        url="jdbc:sasiom://hunding:8591"
```

```
        user="sasdemo"
        password="sasuser"/>

<sql:update dataSource="${class}">
      <%=strSQL %>
</sql:update>

<body style="font-family: verdana, tahoma, arial;">
  <p>The class file has been updated.</p>
</body>
</html>
```

The page uses the `sql:update` tag to execute the constructed SQL code. Note that for a change of pace, the program uses the IOM JDBC driver instead of the SAS/SHARE driver. When the page is displayed in a browser window, all it says is "`The class file has been updated.`" By now, you should be able to combine the two JSTL examples to create a dandy interactive Web site for your application.

Web Archive Files

The examples shown so far have all been extremely simple. In the real world, most Web applications include multiple servlets and/or JavaServer Pages, usually linked to a package of classes. In order to make it easier to deploy complex applications, Sun has introduced the concept of *Web Application Archive* (WAR) files. According to the *Java Servlet Specification 2.2*, a Web application can be a collection of JSP, servlets, HTML, Java classes, and other resources; see http://java.sun.com/j2ee/tutorial/1_3-fcs/doc/WCC3.html for a tutorial. A WAR file is essentially just a regular JAR file containing all of the components necessary to deploy a given Web application.

A Web application can be run from a WAR file directly, or from an unpacked directory with the same structure. The top level of the directory is the *document root* of the application. An example of a document root using Tomcat would be **$TOMCAT_HOME/webapps**. As shown in Display 8.2, the **webapps** directory should contain a subdirectory called **WEB-INF**, which in turn can contain some or all of the following files and directories:

- **web.xml** – the required Web application deployment descriptor (see Example 8.3
- tag library descriptor files (see the following section)
- a **classes** directory that contains any needed servlets, utility classes, and other components
- (optional) a **lib** directory that contains JAR archives of libraries (tag libraries and any utility libraries called by server-side classes)

It is also possible to create package directories in either the document root or the `classes` directory (see the Sun WebApp tutorial). Note that this is pretty much the default Tomcat structure described above; the WAR file mechanism is just a way to package the contents in a fashion analogous to the familiar ZIP or UNIX TAR files.

Building a Web Archive File

It is possible to create the file manually using the JAR utility supplied with the Java SDK (see http://access1.sun.com/techarticles/simple.WAR.html for an example). The following Windows batch command will create a small file called **BBU.war** that contains the JavaServer Page from Example 8.4:

```
C:\j2sdk1.4.2_04\bin\jar cvf BBU.war HelloWorld.jsp
```

The three options to the jar command are:

- c for create file
- v for verbose
- f for the name of the file to create

As discussed in the section on servlets, in order to deploy a WAR file that contains a servlet, it is first necessary to create a **web.xml** application descriptor file. The structure of the deployment descriptor file is fairly complex and will not be described in detail in this book. A good introduction is the Sun tutorial cited previously, as well as various other online sources.[3]

There are several other ways to create a WAR file. Sun suggests using the War task of the Ant portable build tool, which is available from the Apache Software foundation at http://ant.apache.org/. Also, as will be explained in Chapter 9, SAS AppDev Studio includes a component, the webAF Package Wizard, that can easily create the appropriate content from a webAF project.

Web Application Manager

Once the WAR file has been created, it must be uploaded to the server. You can, of course, copy it manually to the server if you have Write permissions to the appropriate directory. However, Tomcat includes an administration utility called the *Web Application Manager* that handles this quite simply via the HTML PUT command to port 8080. Enter the URL http://<server-name>:8080/manager/html into your favorite browser window. You should be prompted for the administrator user name and password. If you have forgotten or do not know the values specified when Tomcat was initially installed, these are set in the configuration file **$CATALINA_HOME/conf/tomcat-users.xml**, as shown in the following sample:

```
<?xml version='1.0' encoding='utf-8'?>
<tomcat-users>
  <role rolename="tomcat"/>
  <role rolename="role1"/>
  <role rolename="manager"/>
  <role rolename="admin"/>
  <user username="tomcat" password="tomcat" roles="tomcat"/>
  <user username="role1" password="tomcat" roles="role1"/>
  <user username="both" password="tomcat"
roles="tomcat,role1"/>
  <user username="admin" password="system"
roles="manager,admin"/>
</tomcat-users>
```

---

[3] James Goodwill, "Using Tomcat: Java Web Applications," O'Reilly & Associates, Inc., 2000-2002. http://www.onjava.com/pub/a/onjava/2001/03/15/tomcat.html.

Also, be careful typing the URL; it's not `manager.html`, but `manager/html` for the Manager application. Something like the following window should appear:

Display 8.6 JWSDP Web Application Manager

Scroll down to the bottom of the screen. Enter the WAR file information as shown and click the **Deploy** button.

Display 8.7 JWSDP Web Application Manager (continued)

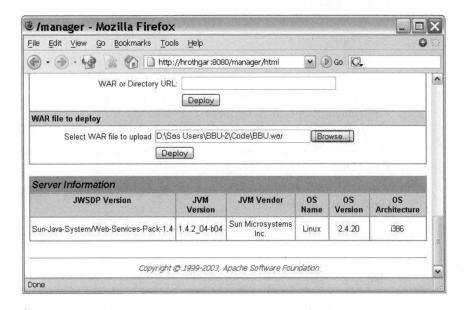

In this case, you should see the message "OK - Deployed application at context path /BBU" in the message text area at the top of the page, indicating that the application has been deployed and loaded into the servlet container and is ready to run.

References

JavaServer Pages

- Bergsten, Hans. 2002. *JavaServer Pages.* 2nd ed. Sebastopol, CA: O'Reilly & Associates.

- Burd, Barry A. 2001. *JSP: JavaServer Pages Developer's Guide.* New York, NY: Hungry Minds, Inc.

- Callaway, Dustin R. R., and Danny Coward. 2001. *Inside Servlets: Server-Side Programming for the Java Platform.* 2nd ed. New York, NY: Addison-Wesley.

- Fields, Duane K., Mark A. Kolb, and Shawn Bayern. 2001. *Web Development with JavaServerPages.* Greenwich, CT: Manning Publications.

- Geary, David M. 2001. *Advanced JavaServer Pages.* Englewood Cliffs, NJ: Prentice Hall PTR.

- Goodwill, James, and Samir Mehta. 2001. *Developing Java Servlets.* Sebastopol, CA: Sams.

- Hall, Marty. 2000. *Core Servlets and JavaServer Pages.* Englewood, Cliffs, NJ: Prentice Hall PTR.

- Hall, Marty. 2001. *More Servlets and JavaServer Pages.* Englewood, Cliffs, NJ: Prentice Hall PTR.

- Hougland, Damon, and Aaron Tavistock. 2000. *Core JSP.* Englewood, Cliffs, NJ: Prentice Hall PTR.

- Hougland, Damon, and Aaron Tavistock. 2001. *Essential JSP for Web Professionals.* Englewood, Cliffs, NJ: Prentice Hall PTR.

- Hunter, Jason, William Crawford, and Paula Ferguson. 2001. *Java Servlet Programming.* 2nd ed. Sebastopol, CA: O'Reilly & Associates.

- Kurniawan, Budi. 2002. *Java for the Web with Servlet, JSP, and EJB: A Developer's Guide to Scalable J2EE Solutions.* Indianapolis, IN: New Riders Publishing.

- Monson-Haefel, Richard. 2001. *Enterprise JavaBeans.* 3rd ed. Sebastopol, CA: O'Reilly & Associates.

- Smith, Dori. 2002. *Java 2 For the World Wide Web.* Berkeley, CA: Peachpit Press.

- Williamson, Alan R. 1999. *Java Servlets: By Example.* Greenwich, CT: Manning.

Custom JSP Tags

- da Silva, Wellington L.S. 2001. *JSP and Tag Libraries for Web Development.* Indianapolis, IN: New Riders Publishing.

- Goodwill, James. 2002. *Mastering JSP Custom Tags and Tag Libraries.* New York, NY: John Wiley & Sons.

- Heaton, Jeff. 2002. *JSTL: JSP Standard Tag Library*. Indianapolis, IN: Sams.

- Shachor, Gal, Adam Chace, and Magnus Rydin. 2001. *JSP Tag Libraries*. Greenwich, CT: Manning Publications.

- Weissinger, A. Keyton. 2002. *Developing JSP Custom Tag Libraries*. Sebastopol, CA: O'Reilly & Associates.

JDBC

- Reese, George. 2000. *Database Programming with JDBC and Java*, 2nd ed. Sebastopol, CA: O'Reilly & Associates.

- Williamson, Alan R. and Ceri Moran. 1997. *Java Database Programming: Servlets and JDBC*. Englewood Cliffs, NJ: Prentice Hall PTR.

- White, Seth, Maydene Fisher, Rick Cattell, Graham Hamilton, and Mark Hapner. 1999. *JDBC API Tutorial and Reference*: *Universal Data Access for the Java 2 Platform*, 2nd ed. Reading, MA: Addison-Wesley.

Links

URL references are current as of the date of publication.

- Configuring Apache to work with Tomcat – http://jakarta.apache.org/tomcat/tomcat-3.3-doc/mod_jk-howto.html

- DBTags download – http://jakarta.apache.org/builds/jakarta-taglibs/releases/dbtags

- Jakarta DBTags Library – http://jakarta.apache.org/taglibs/doc/dbtags-doc

- JavaServer Pages – http://java.sun.com/products/jsp/

- Java Web Applications – http://www.onjava.com/pub/a/onjava/2001/03/15/tomcat.html

- JDBC - http://java.sun.com/products/jdbc

- JDBC Data Access API – http://industry.java.sun.com/products/jdbc/drivers

- JDBC/ODBC Bridge – http://java.sun.com/j2se/1.3/docs/guide/jdbc/getstart/bridge.doc.html

- JSP Tag Libraries – http://java.sun.com/products/jsp/taglibraries.html

- Servlets – http://java.sun.com/products/servlet

- Standard Tag Library – http://java.sun.com/products/jsp/jstl/

- Sun Web Servers – http://wwws.sun.com/software/product_categories/web_servers.html

- Tag Library DTD – http://java.sun.com/dtd/

- Tomcat – http://jakarta.apache.org/tomcat

- Tomcat configuration – http://jakarta.apache.org/tomcat/tomcat-4.0-doc/config

- Web Application Files – http://java.sun.com/webservices/docs/ea2/tutorial/doc/

- WSDP Deploytool – http://java.sun.com/webservices/docs/2.0/ReleaseNotes.html

Chapter 9

Developing Java Server-Side Applications with webAF Software

Getting Started with webAF Software

A quick review: Part 2 showed how to develop CGI server-side applications with the SAS Application Dispatcher and htmSQL. This chapter illustrates a different approach to server-side Web application development, using SAS AppDev Studio 3.1 to create servlets and JavaServer Pages in webAF software. With SAS AppDev Studio software, you can create Java programs that access the features of SAS from within a *thin client* application. This does not necessarily mean that you have to put your users on a diet; a thin client is an application running on a system that does not have SAS installed. Most commonly, this will be via a browser to render Web pages that in turn access SAS data sets and procedures.

webAF is the *Integrated Development Environment* (IDE) for creating Java code in SAS AppDev Studio software. This is not just another pretty interface, however; webAF software also includes a set of customized Java classes that interface with various SAS features. To simplify the process of using these, SAS has provided a collection of wizards to generate Java code and extensive help files and examples of how to use the classes. See http://support.sas.com/rnd/appdev/webAF/ for an overview of webAF software capabilities.[1]

The big operational difference from SAS/IntrNet software, as the preceding chapter pointed out, is that JSP and servlets require a servlet engine running on the server platform in addition to the (optional) Web server software. SAS supplies a copy of the Tomcat servlet engine, along with the Apache HTTP server and the Java JDK. If you are already using versions of Apache, Tomcat, or the JDK that are more recent than those provided with the webAF software, you may be tempted to use these instead. Life is complicated enough already; even if you are an expert at Java Web server configuration, just use the components that SAS recommends. They work more than adequately, they have been tested extensively, and all of the tools and wizards assume the standard versions. Start with a clean system, preferably one with Windows 2000 or XP Professional installed.

All of the examples shown in this chapter assume this standard configuration. According to SAS Tech Support Note SN-013180:

> webAF 3.0 will not run with JDK 1.4.2_05 or later.... At the present time, there is no circumvention other than to revert to JDK 1.4.2_04 or an earlier compatible version. (http://support.sas.com/techsup/unotes/SN/013/013180.html)

In order to get webAF software to work, you must download the older version of the JDK from http://java.sun.com/products/archive/j2se/1.4.2_04/. The important thing to recognize is that whatever version of the JDK you use to create applications with webAF software, they should run fine in more recent versions of the Java Runtime Environment (JRE).[2]

---

[1] In SAS AppDev Studio 3.2, webAF software will be discontinued in favor of the open-source Eclipse development tool. The concepts discussed in this second edition, however, should still apply.
[2] When SAS AppDev Studio 3.2 is released, presumably this limitation will be removed.

The SAS Education Division formerly offered a four-day class entitled "Server-Side Web Application Development Using webAF"; see http://support.sas.com/training/us/crs/ ssjdaf.html. Presently, there are no plans to offer this class after the next release of SAS AppDev Studio software. As always, the SAS AppDev Studio 3.1 Developer's Site at http://support.sas.com/rnd/appdev/ should be consulted for the most current information about documentation and training resources.

The focus in this chapter, as one would expect from a SAS Press book, is on the users' perspective. How, in practice, does this stuff actually work? What tricks of the trade and gotchas are there? Most importantly, how can you use these development tools to add value and utility to your Web pages?

The obvious question at this point is whether it is obligatory to know Java in order to use webAF software; correspondingly, Java programmers would like to know whether they have to learn SAS. The answer in both cases is a resounding yes—and no. While it is possible to use webAF software productively by reading the documentation and following the examples, the experience will be enhanced significantly by acquiring more practice with both Java and SAS (and of course, by reading this book). The SAS AppDev Studio Developer's Site FAQ at http://support.sas.com/rnd/appdev/doc/faq.htm offers links to several online resources on Java.

In contrast to most SAS software, webAF software is platform specific; it is available only for recent versions of the Microsoft Windows operating system: Windows 98 (but not ME), NT4.0, 2000, and XP Professional. Since Java is platform independent, all applets and Web applications created with webAF software will run in any Web browser that supports the Java Plug-in; it is only the IDE that requires the Windows API.

Installing SAS AppDev Studio Software

SAS AppDev Studio software is licensed separately; the code is supplied on a single CD, labeled *SAS AppDev Studio Java Components.*[3] At the time of this writing, the most recent release for SAS AppDev Studio software is the Java Components 3.1.4 update, released in July 2005. This download includes the following components:

- Visual Data Explorer, which is used to view information maps for OLAP cubes and tables (requires SAS BI)

- SAS Web Report Viewer, which is used to access and display reports that are built using SAS Web Report Studio (requires SAS BI)

- additional JSP and swing-based relational and OLAP tables

- prebuilt dialogs, which enable you to interact with data to perform common tasks such as filtering, adding calculated items, defining conditional highlighting rules, ranking, and more

- an expanded API for the Visual Data Explorer component, which will enable you to customize any aspect of the composite component

---

[3] There is also another CD, *SAS AppDev Studio Server Components.* These updates should be applied to the SAS application server (see the following discussion).

The SAS AppDev Studio license requires that Base SAS software is licensed and installed on the development machine. The following discussion of installation procedures is based on the information in the document "SAS AppDev Studio Java Components 3.1.4 Update," which is available at http://www.sas.com/software/distribution/readme/appdevjava314_PROD_/readme_win_108297.pdf.

Depending on the installation, in addition to the usual *InstallShield* prompts, the user may be asked several questions that require some specialized knowledge:

1. "If you do not have a Web server, setup can install the Apache WebServer at this time. Do you want to launch the install?"

You do not need to install a Web server on your local workstation in order to use SAS AppDev Studio software. It can be very useful, however, to be able to test your Web pages as you develop them. Most Web server administrators are reluctant to let users install test versions of user pages on a production server, so if you do not have access to a development server you might want to install one on the same system as the client. There are several choices. SAS AppDev Studio software comes with a Windows version of Apache that should work acceptably.

2. "What is the root URL for your Web server?"

The default value for this is the network computer name of your workstation; there is no reason to change this, even if you decided in the previous step not to install a Web server.

3. "Setup cannot locate the Java Runtime Environment. Do you want setup to install the JRE at this time?"

As noted previously, SAS AppDev Studio software can be very picky about which version of Java is installed; it is a good idea to use the recommended one, currently j2sdk1.4.2_04.[4] If you have a newer version of the JRE, you will have to uninstall it (select **Add/Remove Programs** from the Windows Control Panel) before the SAS AppDev Studio installation can be completed.

If you get the preceding message, you must reply "Yes" or the process will abort. If you have enough disk space in the specified directory, install all the options. Otherwise, you can deselect **Java Sources**, **Old Native Interface Header Files** and **Demos** in order to reduce the size of the installation.

One further installation note: SAS AppDev Studio 3.1 requires installing a set of server-side upgrades. (These go on the SAS server, not the Tomcat server.) The upgrades are on a separate disk, supplied with the installation media. SAS recommends that you use the Software Navigator to install them. The program will install the necessary catalog updates for SAS AppDev Studio 3.1.

---

[4] The Sun Developer Network site at http://java.sun.com/javase/downloads/ notes that this Java release has "completed the end of life process" and therefore is no longer supported. You can still download the older versions by going to the archive site and selecting JDK 1.4.2_04, which seems to work correctly.

The webAF Integrated Development Environment

Once SAS AppDev Studio software has been installed (and updated) you should be able to start webAF software by selecting **Start ▶ Programs ▶ SAS AppDev Studio ▶ webAF**. A splash screen appears and displays a series of startup messages as the various Java components are initialized; this may take a while, depending on the operating system and the speed of your processor.

Like most IDE software, webAF software can be pretty daunting at first glance. As shown in Display 9.1, the default settings include three sets of windows with tabs, a menu bar, and six (out of eight possible) toolbars:

Display 9.1 webAF Software

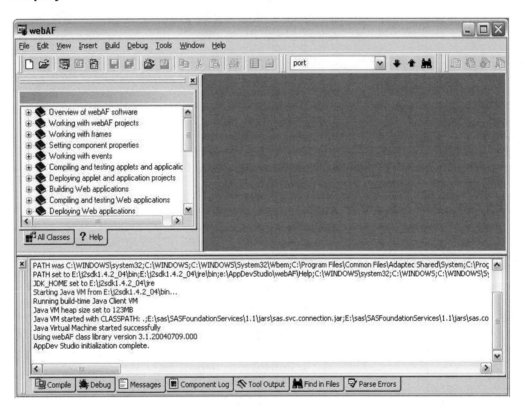

Actually, it's not all that hard to figure out. The view displays the following windows:

- The Help table of contents initially appears in the Project Navigator window at the upper left; the other tab labeled All Classes shows the default Java packages available until a project is loaded.

- The Code window at the upper right is initially blank; this is used for editing text when a project is loaded.

- The Message window (across the bottom) has seven tabs for selecting different types of output to display.

In addition to these windows, the **View** menu also enables you to toggle on and off various components, such as the Status Bar, Debug Variables window and Call Stack window. You probably want to start with the Status Bar on (where it says "For Help Press F1" in the bottom left corner of the previous example) and the other two off.

You can also choose which tool bars to display by selecting them from the following list (on the **View ▶ Toolbar** menu):

Table 9.1 webAF Toolbars

Toolbar	Default Buttons and Text Boxes, with Keyboard Shortcuts
Standard	New (Ctrl+N) Open (Ctrl+O) New Project (Ctrl+R) New Frame (Ctrl+M) New Java Source (Ctrl+J) Save (Ctrl+S) Save All Open Remote SCL Save SCL Remotely Copy (Ctrl+C) Cut (Ctrl+X) Paste (Ctrl+V) Print (Ctrl+P) Component property sheet Component customizer
Find	Find string text box Find next Find previous Find
Debug	Go (F5) Break Restart Quit (Shift+F5) Step into (F11) Step over (F10) Step out (Shift+F11) Run to cursor Toggle breakpoint (F9) Toggle breakpoint enable (Shift+F9) Remove all breakpoints Toggle debug variables window Toggle call stack window

Toolbar	Default Buttons and Text Boxes, with Keyboard Shortcuts
Alignment	Align left Align horizontally Align right Align top Align vertically Align bottom Same width Same height Same size Even horizontal spacing Even vertical spacing Color Font Border Dot grid Line grid No grid
Methods and Fields	Methods text box Fields text box
Build	Compile file (Ctrl+F7) Build (F7) Stop build Rebuild all Execute (Ctrl+F5) Execute in browser Go (F5) Toggle breakpoint
SCL Debug	SCL Command Line text box

You probably want to start by deselecting the Find, Alignment, Methods and Fields and Breakpoint toolbars. You can turn them back on if you decide you want to have these functions handy, but since they provide some special-purpose functions, you probably do not need them initially.

The **Tools** menu lists following choices:

- **Edit Component Palette** – used to add or remove controls from toolbar
- **Event Manager** (deselected until a project is opened) – used to create, modify, or delete event handlers that specify interactions between components
- **Handle Event** (deselected until a project is opened)
- **Portlet Manager** (deselected until a portlet project is opened) – new in SAS AppDev Studio 3.1
- **Register Connections** – used to define data sources
- **Register Layout Managers** – used to add or remove applet format containers
- **JSASNetCopy Control Panel** – used to manage Java applet JARS and certificates
- **Update JSASNetCopy .war** – automatic client-side installation of Java libraries

- **Wizards** – utilities used to automate tasks, including the DataBean, EJB Generation, and the Information Bean wizards for creating different types of Java Beans, and the Package Wizard for creating Web archive (WAR) files
- **Reports** – used to create Property Links, Model/View Attachments, Events and Diagnostic reports
- **Services** – used to start and stop the SAS AppDev Studio Middleware server and the Java Web (Tomcat) server
- **User Tools** – used to start and stop other services, such as J2EE and the Cloudscape database server
- **Source Control** – used to configure the behavior of projects under source code control (SCC), new in SAS AppDev Studio 3.1
- **Options** – used to select and modify webAF display options
- **Customize** – used to customize commands, toolbars, and keyboard shortcuts

All of these interface elements—menus, windows and toolbars—can be extensively customized for user preferences. The "Customizing the webAF Environment" topic in the webAF online Help explains how to rearrange the IDE to suit your specific needs. (In general, the menus available from the Project Navigator **Help** tab provide a great deal of information about the visual development environment.)

Creating a JSP Project

webAF software, like Microsoft Visual Studio and many other development environments, depends on the concept of a software *project*. This is simply a collection of files relating to some specific application. SAS uses a project file with the extension .afx to keep track of this collection. Prior to SAS AppDev Studio 3.1, it was necessary to create these directories manually. In the current release, the WebApp Project Wizard (see Display 9.2) takes care of this automatically.

To create a new Web application in webAF software, select **File ▶ New**. From the **Projects** tab, select **Web Application Project**. You should see something like the resulting screen:

Display 9.2 Creating a New JSP Project

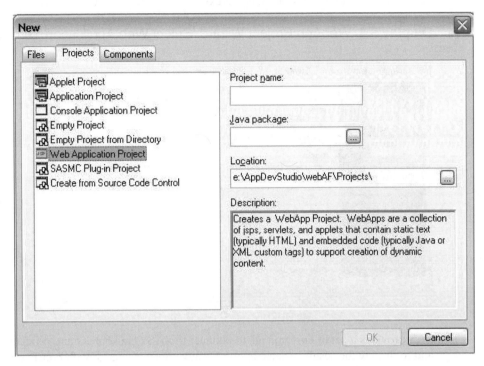

Three text boxes show up: **Project name, Java package** and **Location**. The only one of these that you must specify is the project name—in this case `Examples`. Leave the second text field blank. The result is that the WebApp Project Wizard does not create a hierarchy of directories to store Java class files (see the webAF Help topic "Assigning a Package Name" for more details). The **Location** field will automatically fill in with the default SAS AppDev Studio Project directory name; again, this is probably what you want to do.

For this example, the WebApp Project Wizard then proceeds through seven steps, the first of which is shown in Display 9.3. The WebApp Project Wizard lists only six steps because it assumes that you want to create a new application from scratch, rather than using the built-in templates as in this example.

Display 9.3 WebApp Project Wizard: Select Web Application Location

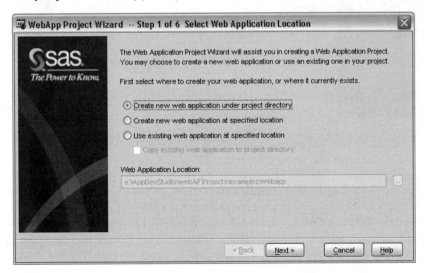

The default selection, **Create new Web application under project directory**, is almost certainly what you want to do. Click **Next**.

Display 9.4 WebApp Project Wizard: Select Web Application Template

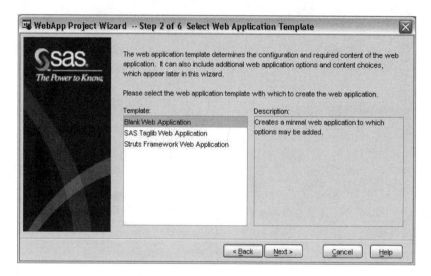

SAS provides a set of custom tags to connect to SAS data sources and procedure output and to display the results. (See Chapter 8, "Java Servlets and JavaServer Pages," for an explanation of custom tags in JavaServer Pages.) Apache Struts is an open source framework for building Java Web applications. The Struts homepage is available at http://struts.apache.org/. Finally, the WebApp Project Wizard can create a blank Web application.

Display 9.5 WebApp Project Wizard: Select Web Application Options

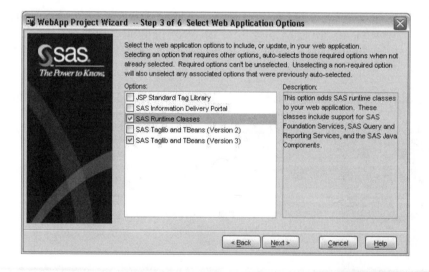

Selecting **SAS Taglib Web Application** requires a choice of tag libraries. Two sets of SAS Appdev Studio custom tags are provided: Taglib Version 2 and Taglib Version 3. These are shown in Display 9.5. (A *TBean* is a SAS *TransformationBean* Java class. Don't worry about how to implement a TBean; the tag library will take care of it for you.) Depending on what you want to do, you can use either or both Version 2 and Version 3 custom tags, along with the JSP standard tag library. The examples below show the use of both sets.

The default selection is to add the Version 3 libraries, along with the standard SAS run-time classes. (Once the project has been built, the Java class libraries can be viewed from the Project Navigator.)

Display 9.6 WebApp Project Wizard: Select Web Application Initial Content

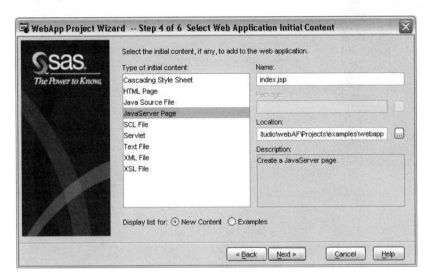

The default for the initial content of a Web application is the JavaServer Page **index.jsp**. Choosing the **Examples** radio button will create a prewritten Web application from among a set of choices. For this example, selecting **New Content** will just create a blank page.

Display 9.7 WebApp Project Wizard: Select JavaServer Page File Template

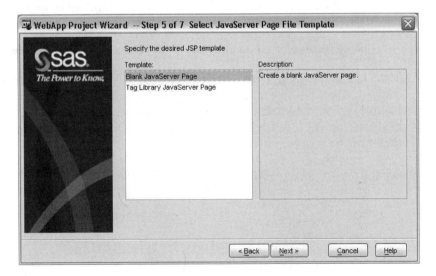

For this example, select **Blank JavaServer Page**. The next screen provides five text boxes:

Display 9.8 WebApp Project Wizard: Specify WebApp Project Options

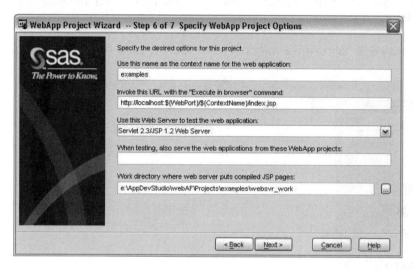

These text boxes call for the following information:

1. Context name: in this case, `examples,` sets the value of the `ContextName` project variable.
2. URL: the resource location used to test the JavaServer Page; `WebPort` defaults to 8082, whereas `ContextName` is defined in the preceding text box.

The SAS AppDev Studio software installation procedure creates two separate Tomcat installations. One starts Tomcat on port 8084 from the desktop shortcut **Start ▶ Programs ▶ SAS AppDev Studio ▶ Services ▶ Start Java Web Server**. This link is not to the standard Tomcat **startup.bat** script from the Jakarta project but instead points to the following file:

  \AppDevStudio\bin\Launcher.exe -jdk server StartWebServer.cfg

The Launcher program is supplied by SAS specifically for SAS AppDev Studio software. There is also a shortcut on the **Programs** menu to stop the server, which logically enough calls the Launcher with **StopWebServer.cfg**. The **StartWebServer.cfg** file is shown in Display 9.9 for illustrative purposes only; it should not be necessary to modify this file.

Display 9.9 Web Server Startup Configuration File

```
// start java webserver
-Morg.apache.catalina.startup.Bootstrap
-TJava Web Server
-Dcatalina.base=e:\AppDevStudio\Java\WebServer\sc23
-Dcatalina.home=e:\AppDevStudio\Java\WebServer\sc23
-Djava.io.tmpdir=e:\AppDevStudio\Java\WebServer\sc23\temp
-Djava.class.path=e:\AppDevStudio\Java\WebServer\sc23\bin\
                  bootstrap.jar;
       <JDKPATH>\lib\tools.jar
Start -config conf\server-sasweb.xml
```

Note that the server will start only if the Java Runtime Environment 1.4.1 is installed and the correct registry key is found. For more information on the Launcher program and the run-time flags shown, see the webAF Help topic "Working with AppDev Studio Service Configuration Files."

The other servlet engine is started on port 8082 from within webAF software when you select **Tools ▶ Services ▶ Start Java Web Server** from the main webAF menu. This server is stopped by selecting **Stop Java Web Server**. These shortcuts control a separate servlet engine for testing applets and applications from within webAF software; the server configuration file for this can be found here:

`\AppDevStudio\WebAppdev\conf\server.xml`

3. Web server: the default Java Web server is the Tomcat servlet engine on port 8082
4. Additional Web applications: usually left blank
5. Work directory: the directory where Java source and class files are stored

The servlet engine creates a Java source file and a compiled class file from the JSP; these two files are saved in the **work** directory. The servlet engine loads the class file and executes it, sending the generated HTML back to the client browser. If the project name is examples as illustrated previously, the compiled programs are stored by default in the following directory:

`C:\AppDevStudio\webAF\Projects\examples\websvr_work\examples`

The first time the JSP page is loaded, a new Java program is generated and saved in this directory. This has two important implications. First, the servlet engine cannot always tell when the class file has changed and should be reloaded. One of the most common and frustrating problems that JSP developers experience is changing the JSP code, opening the page in a browser window, and seeing the result of the previous compilation unchanged.

If this happens to you (and it will), go to the **work** directory and delete all the files there. Then the next time you open a page, it will seem to take forever to load. This is because the servlet engine has to go through the whole three-step process of generating the servlet source code, compiling it into a class file and loading it for execution. The second and subsequent executions of the same program are much quicker, because the class file has already been loaded. (For production servers, it is possible to precompile JSP libraries.)

The second implication is that any compiler error messages that show up will reference the generated Java source code file; this is the program stored in the **work** directory. To find the error, you need to open the Java file to find the line number of the error. Don't try to look for it in the JSP file.

Display 9.10 WebApp Project Wizard: Summary

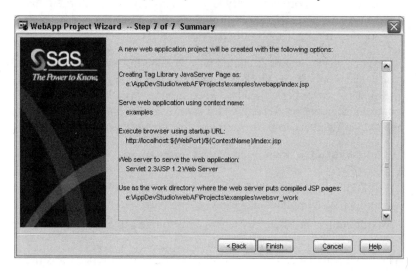

Click **Finish** to create the project using the options shown. Note that SAS AppDev Studio software is doing a lot of work for you, so this may take a while, even on a fast machine.

A Simple JSP Example

Once the `examples` project has been created, the JSP/Servlet window should open to the **Source** tab. For illustration purposes, in Example 8.3 from the previous chapter, the `HelloWorld` JavaServer Page can be pasted into the JSP source window, with the result shown in Display 9.11.

Display 9.11 Hello World Example: Source Tab

```
1 <!DOCTYPE html PUBLIC "-//W3C//DTD XHTML 1.0 Transitional//EN"
2    "http://www.w3.org/TR/xhtml1/DTD/xhtml1-transitional.dtd">
3 <html xmlns="http://www.w3.org/1999/xhtml">
4 <!-- Simple JavaServer Page Example -->
5 <head>
6    <title>JSP Examples</title>
7    <style type="text/css">
8       h1   { color: blue; }
9       body { font-family: helvetica, arial; }
10   </style>
11 </head>
12 <body>
13    <h1>Simple JavaServer Page</h1>
14    <% out.print("Hello World!"); %>
15    The time now is <%= new java.util.Date() %>
16 </body>
17 </html>
18
```

It is also possible, if not very useful, to select the **Visuals** tab to see what this page looks like in webAF software.

Display 9.12 Hello World Example: Visuals Tab

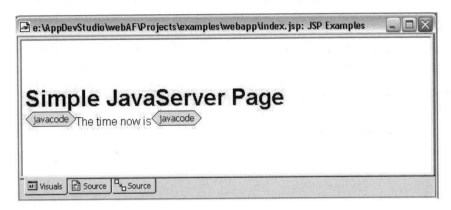

In addition to the JSP, the wizard will insert a file called `web.xml`. This file is used by the servlet engine to manage servlets; see the discussion in Chapter 8. For this example, you can just ignore this file.

To see what this page looks like when displayed by the servlet engine, start the servlet engine from the webAF menu by selecting **Tools ▶ Services ▶ Start Java Web Server**, as explained previously. Once the server is running on port 8082, select **Build ▶ Execute in browser ▶ Internet Explore**. The output that appears in the browser window should look exactly the same as Display 8.3 in the preceding chapter.

Another way to start the Web server is by pressing the F5 function key or selecting **Go** from the Build toolbar. This will open the debug window, where useful messages may appear. Note too that if you close webAF software, the WebAppDev Web server on port 8082 is also shut down automatically.

Of course, using webAF software to display previously constructed pages is not very sensible; the real value of webAF software is in creating JSP files directly, using the webAF integrated development environment.

Building Forms with webAF Software

The following example is a literal translation into JSP of the CGI temperature conversion calculator from Chapter 5, "Web Applications Programming." (You might want to refer back to the program `convert.pl` to see the differences.) Of course JSP works more like JavaScript than Perl; the page shown in the following example is basically just a regular XHTML page, with a few added elements to render it dynamic.

The structure of a JavaServer Page that contains a form can best be understood as the combination of three related pieces:

- static XHTML template code
- one or more Java scriptlets
- HTML form code containing Java expressions

Creating a New JavaServer Page

Building the temperature conversion calculator in webAF software starts with creating a new project or adding a new page to an existing one. To create a new page in an open project, just select **File ▶ New** from the main menu. Enter a new file name, in this case **F2C.jsp** (for Fahrenheit-to-Centigrade converter). This time you want to create a page using the SAS tag library, so when prompted by the wizard (see Display 9.7) select **Tag Library JavaServer Page**. The JSP/Servlet window should open as in Display 9.13 with a new blank page.

Display 9.13 Creating a New JavaServer Page

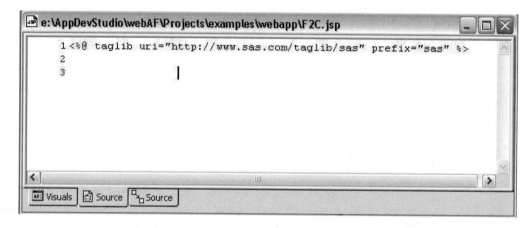

The first line is the link to the SAS AppDev Studio Version 3 tag library which, as we will see, is used by webAF software to create form elements.

```
%@taglib URI="http://www.sas.com/taglib/sas" prefix="sas"%
```

Note that the URI attribute looks a lot like an anchor tag. You do not need to connect to the Internet in order to display a page containing this tag, however; the local Java Web server understands this reference not as a link, but instead as an identifier. This is just the name of the resource, not its location.

To create the page, first insert the static XHTML portion of the code.

Example 9.1 JSP Header

```
<% @taglib URL="http://www.sas.com/taglib/sas" prefix="sas"%>

<!DOCTYPE html PUBLIC "-//W3C//DTD XHTML 1.0 Transitional//EN"
    "http://www.w3.org/TR/xhtml1/DTD/xhtml1-transitional.dtd">
<html xmlns="http://www.w3.org/1999/xhtml">
<head>
  <title>JSP Examples</title>
</head>
<body>
<h1 style="color: blue">Temperature Conversion Calculator</h1>
    [Java scriptlet comes next]
    [HTML form code follows]
</body>
</html>
```

Adding Styles to JavaServer Pages

The first six lines of the code are the same for all the examples in this chapter. It makes sense, then, to save them as a separate file and just include them with a JSP directive on each page. In this way, any changes to the XHTML header can be made in one place. Saving the first six lines as file **header.html** allows them to be included in the JSP with the following statement:

```
<%@ include file="header.html" %>
```

The last two lines could also be saved as file **footer.html** and inserted at the bottom of each JavaServer Page. Although in this example there are only two lines, it would be possible to add an image tag like "Powered by SAS" to the bottom of every page just by modifying this one file.

As Chapter 8 indicated, the JSP `include` directive does not automatically pick up changes in the included file. The servlet code has to be regenerated to update the page. While this is fine for most applications, an alternative for files that are likely to be frequently modified is to use the `jsp:include` action, which will include the file at the time of the client request, at the cost of a little more processing time.

In order to create a common style for all the Web pages at your site, it is convenient to use a style sheet. A link to an external CSS can easily be added to the **header.html** file. SAS supplies several style sheets in the **webapp\assets** subdirectory of the project folder. Including a link to one or more of these style sheets will provide output in the SAS style. The files are generated by the WebApp Project Wizard. Depending on the type of project desired, one or more of the following may appear:

- **sasStyle.css**
- **sasads.css**
- **menu.css**
- **sasads_formElements.css**
- **sasComponents.css**
- **sasSimpleCalculatedItemSelector.css**

In order to figure out which style sheet applies to a given component, you need to review the SAS documentation on using the supplied styles; see "Cascading Style Sheets, Images, and SAS TransformationBeans" at http://support.sas.com/rnd/appdev/V2/webAF/server/usingstyles.htm and "TransformationBean Style Sheet Reference" at http://support.sas.com/rnd/appdev/V2/webAF/server/adsstyles.htm.

Writing the Java Scriptlet

The second step is the Java code to calculate the temperature conversion. The following scriptlet is based on the Perl script shown in Chapter 5, with the big difference being that while Perl has no variable types to speak of, Java is a strongly typed language. Most of the changes are just to make sure the Java code will compile with no exceptions.

Example 9.2 Java Temperature Conversion Scriptlet

```
<%      // Java scriptlet to calculate temperature conversion
   String temp = request.getParameter("input");
   String type = request.getParameter("convert");
   String result = new String();

   // make sure that we have values for the parameters
   if (null != temp && null != type)
   {
        double dt = new Double(temp).doubleValue();
   if (type.charAt(0) == 'F')
        result = String.valueOf(5.0 * (dt - 32.0)/9.0);
   else if (type.charAt(0) == 'C')
        result = String.valueOf((9.0 * dt) / 5.0 + 32.0);
   }
%>
```

The Java code is inserted into the page enclosed in <% and %> scriptlet tags. (A scriptlet contains Java code that is executed when the JSP is invoked; see Chapter 8 for more information about writing JavaServer Pages.)

To get the values appended to the URL, the program uses the getParameter() method of the HTTP request object. The parameter values for the input temperature and the conversion type are assigned to string variables in Java (temp and type). The two possible values for type are Fahrenheit to Centigrade and Centigrade to Fahrenheit so we can use the first character of each ('F' or 'C') to decide on the appropriate formula. Then the temperature is converted to a double precision floating point value (so you can do arithmetic on it), used to compute the answer and assigned to the string result.

Building the XHTML Form

Finally, a form is used both to collect and display the information. (In practice, you might want to have one page for input and another for results, but for illustrative purposes it is simpler to show one page for both.) The form makes use of the sas custom tag library. Remember that webAF is an IDE. By dragging and dropping widgets from the component palette onto the source page, you can generate custom tags automatically.

Display 9.14 The Form Element Component Palette

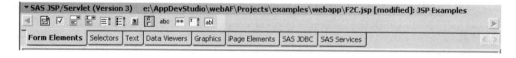

The tool box displayed on the **Form Elements** tab at the top of the JSP/Servlet window has 12 controls available. The following icons are shown in Display 9.14 from left to right:

- sas:Form
- sas:CheckBox
- sas:ChoiceBox
- sas:ComboBoxView

- sas:ListBox
- sas:ListBoxView
- sas:PushButton
- sas:Radio
- sas:Hidden
- sas:Password
- sas:TextArea
- sas:TextEntry

For a detailed description of these tools, including the attributes that can be specified for each, see http://support.sas.com/rnd/gendoc/bi/api/Components/taglibs/sas/overview.html.

This example makes use of three of these form controls: TextEntry, Radio and PushButton. The webAF form controls are custom tags that are based on SAS *TransformationBeans.* TransformationBeans can be used to create HTML widgets, offering additional flexibility over the standard HTML form controls. The whole topic of *InformationBeans* in webAF software is described in detail later in this chapter. For the moment it is worth noting that the tag handler class for the `sas:TextEntry` tag is described in the tag library API under

```
com.sas.taglib.servlet.beans. TextEntryTag.
```

It is possible to drag and drop these controls either to the **Visuals** tab or the **Source** tab in the IDE. The former does not work very well for JavaServer Pages, since there is no HTML as such, so at this point just use the **Source** tab for code development. The **Visuals** tab is useful when formatting your components within a table, since it provides the table layout view.

Dragging a `TextEntry` control to the **Source** tab and dropping it on the page inserts the following custom tag, where the ID attribute specifies the name of the object:

```
<sas:TextEntry id="textEntry1" />
```

This is an okay start, but what we actually want looks more like this:

```
<sas:TextEntry
  id="input"
  prolog="<strong>Enter a temperature and select a conversion
      type: </strong>"
  size="8"
  text="<%=temp%>" />
```

Naming the tag `input` is easier to remember than `textEntry1`. The `prolog` attribute specifies the text that will appear on the page before the text box. (As you might guess, there is also an `epilog` attribute for text to appear after the box.) The width of the text box is specified by the `size` attribute. Finally, the initial text to appear in the box is specified by the Java expression `<%= temp %>`, which is the value of the class variable `temp` defined in the scriptlet code. Initially this will be null, so the text box is empty. Once an input temperature is specified, the resulting parameter value will be inserted into the box.

The other control required on the form is a radio button to select the conversion type. Dropping a Radio widget on the **Source** tab results in the following:

```
<sas:Radio id="radio1" model="" />.
```

Again, not a bad start, but what is really needed is this:

```
<sas:Radio
    id="convert"
    selectedItem="<%=type%>">
    Fahrenheit to Centigrade
    Centigrade to Fahrenheit
</sas:Radio>
```

The object name is now `convert` rather than `radio1`. Two radio buttons are created, with the labels indicated; note that there are no quotation marks around the labels. The expression `<%=type%>` used as the value of the `selectedItem` attribute indicates that if the value of the `type` parameter is equal to one of the two labels (as it will be on the second and subsequent reference to the page), that button is checked. Otherwise, when the page is displayed, neither radio button would be checked and consequently it would be impossible to know which conversion had been executed. (It is probably a good idea to include a declaration for `type` at the top of the page, specifying a default value, to make sure that it is initialized.)

The other two controls on the page are a text box called `output` to display the result of the calculation and a push button to submit the form. Putting it all together results in the following page:

Example 9.3 JSP Temperature Conversion Calculator

```
<%-- JSP Temperature Conversion Calculator --%>
<%@ include file="header.html" %>
<%      // Java scriptlet to calculate temperature
  String temp = request.getParameter("input");
  String type = request.getParameter("convert");
  String result = new String();

  // make sure that we have values for the parameters
  if (null != temp && null != type)
  {
      double dt = new Double(temp).doubleValue();
  if (type.charAt(0) == 'F')
      result = String.valueOf(5.0 * (dt - 32.0)/9.0);
  else if (type.charAt(0) == 'C')
      result = String.valueOf((9.0 * dt) / 5.0 + 32.0);
  }
%>

<h1 style="color: blue">Temperature Conversion Calculator</h1>

<sas:Form id="calculator" action="F2C.jsp" method="get">
  <p><sas:TextEntry id="input"
      prolog="<strong>Enter a temperature and select a
          conversion type: </strong>"
      size="8" text="<%=temp%>" /></p>
```

```
<p><sas:Radio
    id="convert" selectedItem="<%=type%>">
        Fahrenheit to Centigrade
        Centigrade to Fahrenheit
</sas:Radio></p>

<p><sas:TextEntry
    id="output"
    prolog="<strong>Result: </strong>"
    size="8" text="<%=result%>" /></p>

<sas:PushButton
    id="submit"  text="Submit" />

</sas:Form>

<%@ include file="footer.html" %>
```

The requested `action` (when the **Submit** button is clicked) is to redisplay the page, with the parameter values appended to the URL. The form's `method` attribute has been set to `get` so that the parameters will be visible; if this is not desired, specify `post` instead.

This program was constructed just by fitting together the three pieces described previously. Referencing this page in a browser results in the following page (don't forget to start the server!):

Display 9.15 JSP Temperature Conversion Calculator Output

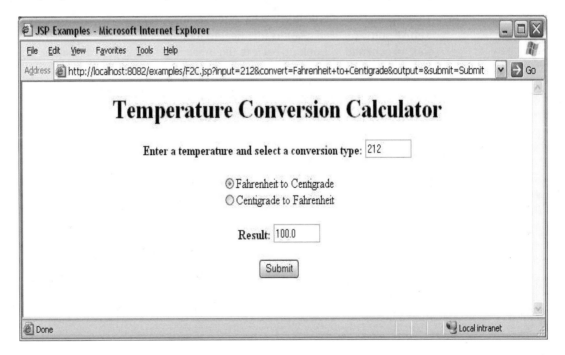

The URL referencing this page includes *name-value* pairs for four parameters:

1. `input` (text box)
2. `convert` (radio buttons)
3. `output` (text box with null value)
4. `submit` (push button)

The following values are shown in the example (white space inserted for clarity):

```
http://localhost:8082/examples/F2C.jsp?
   input=212&
   convert=Fahrenheit+to+Centigrade&
   output=&
   submit=Submit
```

Blank spaces in the values have been replaced with + signs; otherwise they appear as you would expect. The scriptlet code has read the input text and radio controls and calculated the result shown.

Attributes and Models

Before leaving this example, it is worthwhile to look at one minor modification to the radio button. Right now, as shown in the previous output, the value of the `convert` parameter is the label of the button. Using a `model` attribute on the Radio custom tag, it is possible to pass an index value for the button (1 or 2) rather than the whole string.

The explanation of how this works sheds some useful light on how models work in webAF software. The result is identical to the preceding example, but it takes more JSP code to generate it. Looking back to the preceding example, note that the tag for the radio button is:

```
<sas:Radio id="convert"
    selectedItem="<%=type%>">
    Fahrenheit to Centigrade
    Centigrade to Fahrenheit
</sas:Radio>
```

It is also possible to achieve the same result with:

```
<sas:Radio id="convert" model="values"
    descriptionModel="labels"
    selectedIndex="<%=checked%>" />
```

where `values` and `labels` code for the index values (1 and 2) and the button labels. These are JSP attributes associated with the *page context*. In order to implement this, it is necessary to modify the scriptlet, as shown in the following example (the added code is shown in bold):

Example 9.4 Scriptlet with pageContext Attributes

```
<%@ page import="com.sas.collection.OrderedCollection" %>

<%      // Java scriptlet to calculate temperature
   String temp = request.getParameter("input");
   String type = request.getParameter("convert");
   String result = new String();

   // default is 1st radio button
   int checked = 0;

   // add the labels for the radio buttons to the page context
   pageContext.setAttribute
       ( "values",  new OrderedCollection("F,C") );
   pageContext.setAttribute
       ( "labels",  new OrderedCollection
       ("Fahrenheit to Centigrade, Centigrade to Fahrenheit"));

   if (null != temp && null != type)
   {
       double dt = new Double(temp).doubleValue();
       if (type.charAt(0) == 'F')
       {
              result = String.valueOf(5.0 * (dt - 32.0)/9.0);

       }
   else if (type.charAt(0) == 'C')
       {
              result = String.valueOf((9.0 * dt) / 5.0 + 32.0);

              checked = 1;
       }
   }
%>
```

The `values` and `labels` associated with the model and `descriptionModel` attributes are `OrderedCollection` objects. This class is added to the class path with the import statement in the page directive at the top. The constructor for the collection takes a comma-delimited string as an argument; here the `values` collection includes "F" and "C", and the `labels` collection the type of conversion desired. (Do not separate each item with quotation marks; just use commas and surround the string with double quotation marks.)

The statement `pageContext.setAttribute` takes two arguments: the name of an attribute, and an object, in this case an `OrderedCollection`. The result is to associate the name with the object in the page context, and add the name-value pair to the list of available attributes. Consequently, when the term `values` is used in the model attribute of the radio button, it is understood to mean the collection object, that is, the strings "F" and "C".

Finally, the original version used the tag attribute `selectedItem="<%=type%>"`; the revised code has `SelectedIndex="<%=checked%>"`. The selected *item* must be a string, such as the button label `Fahrenheit to Centigrade`. In contrast, the selected *index* has to be an integer. The radio buttons are numbered starting with 0. Consequently, it is necessary to add a new integer variable called `checked` to the scriptlet. The default value is 0, so that in the absence of any other value of `type` the first button is selected. However, if, `type="2"`, then `checked =1` and the second button is selected.

The output of this new code, as noted previously, is identical to the previous version. The HTML source code produced is different, however (white space added for clarity):

Display 9.16 Page Attribute Output HTML Source

```
<?xml version="1.0"?>
<!DOCTYPE html PUBLIC "-//W3C//DTD XHTML 1.0 Transitional//EN"
   "http://www.w3.org/TR/xhtml1/DTD/xhtml1-transitional.dtd">
<html xmlns="http://www.w3.org/1999/xhtml">
<head>
   <title>JSP Examples</title>
</head>
<body style="text-align: center; ">

<h1 style="color: blue">Temperature Conversion Calculator</h1>

<form name="calculator" method="get" action="F2C.jsp">

<p><strong>Enter a temperature and select a conversion type:
</strong>
<input type="text" name="input" value="212" size="8" /></p>

<p><input type="radio" name="convert" value="F"
checked="checked">
Fahrenheit to Centigrade</input><br/>
<input type="radio" name="convert" value="C">
Centigrade to Fahrenheit</input></p>

<p><strong>Result: </strong>
<input type="text" name="output" value="100.0" size="8" /></p>

<input type="submit" name="submit" value="Submit" />

</form>

</body>
</html>
```

The generated HTML form is shown in bold. The custom tags have each been rendered as form controls; the values of the attributes supplied are converted into the appropriate HTML code. While this is a trivial example, it is instructive to see how the webAF controls can be modified with a few well-chosen lines of Java program code.

Connecting to SAS Data

The first step for using Remote Data Services in webAF software is to register one or more connections. The two possible choices for a PC-to-PC or PC-to-UNIX connection are either to a SAS/CONNECT spawner or to a SAS/IOM job spawner.

A local host connection is simply a special case of one of these. The neat thing about TCP/IP is that the protocols do not care whether the other end of the pipe is on the same computer, the local host, or on the other side of the world, via the Internet; the communication procedures are the same.

Many servers do not allow Telnet connections because of the inherent security problems with this older technology; consequently SAS recommends using a connection to a spawner.[5] (Chapter 1, "SAS and the Internet," discusses TCP/IP connections in general, including Telnet, while Chapter 4, "Remote Access to SAS," includes detailed examples of how to start spawners on Windows or UNIX platforms.)

Registering Connections

To view the connections available, or to create a new one, select **Tools ▶ Register Connections** from the webAF main menu. (It is not necessary to have a project open; a connection can apply to any webAF project.) Something like the following list of persisted connections should appear:

Display 9.17 Register Connections Tool

When you start webAF software for the first time, you will see only one connection: the default one. This is actually just the skeleton of a connection; some additional work is necessary to implement it. In addition, if you are using a remote SAS data server, as this book recommends, you need to be aware of the issues involved in communicating with object spawners on different platforms. The following example shows a connection to a Linux host. The procedure for setting up and testing a connection to a Windows IOM server is the same, except that the connection is to port 8591, the workspace server, rather than the object spawner.

Click the connection name to select it. Selecting **Edit** should display a window something like the one shown in Display 9.18.

---

[5] The SAS Help and Documentation topic "Help on SAS Software Products: SAS/CONNECT Software" has a section called "Why Use a SAS/CONNECT Spawner" that goes into more detail on the advantages of signing on to a spawner.

Display 9.18 Edit Connection

The connection shown uses the IOM server configuration to host name hygelac on port number 5308. The latter is the port for a connection to an object spawner on a Linux host. (See Chapter 10, "Using the SAS Open Metadata Architecture with the Integrated Object Model," for an explanation of how to start an IOM server.)

Assuming that the spawner service is running, the only change you need to make is to add a connection name in the text box at the top. A useful convention is to use the server name—in this case, to label a connection to hygelac, call the connection "hygelac."

To test the connection, click the **Test** tab, and then click **Check Connection**. The system will attempt to connect to the specified host. The connection test window should display the following message if the connection test is successful:

Display 9.19 Successful Connection to Local Host

```
Threaded connection test starting...
NOTE: PROCEDURE PRINTTO used (Total process time):
      real time          0.00 seconds
      cpu time           0.00 seconds

Connection Success!!
```

At the same time, a terminal window on the host will display the SAS log of the connection:

Display 9.20 Successful Connection to Remote Host

```
NOTE: Copyright (c) 2002-2003 by SAS Institute Inc., Cary, NC,
      USA.
NOTE: SAS (r) 9.1 (TS1M3)
      Licensed to FREDERICK PRATTER, Site 0041553004.
NOTE: This session is executing on the Linux 2.4.20-8 platform.

NOTE: SAS initialization used:
      real time           0.11  seconds
      cpu time            0.00  seconds

   Inherited client connection (1) for user sas.
   Peer IP address and port are 192.168.2.100:1469.
   Client connection (1) closed.

NOTE: The SAS System used:
      real time           2.05  seconds
      cpu time            0.08  seconds
```

Once you can get this test script to work, you should be able to create persistent connections using these parameters. There are a number of good reasons why this test might fail, however. The most likely is that you did not remember to start the spawner first. Check the second tab, labeled **Prompts,** and make sure that the `login` and `password` prompts are spelled correctly and that the `command` prompt is specified correctly. If you are getting Java exceptions when you run the connection test, make sure all of the required updates are in place.

Using the Model-View-Controller Architecture in webAF Software

The real utility of webAF software is that it provides a set of reusable classes for connecting to SAS data. The provided templates make it relatively simple to create dynamic Web pages to view and edit the content of SAS data sets. These templates are available from the WebApp Project Wizard, at Step 4 on the Select Web Application Initial Content screen as shown in the following display. This is the same screen as in Display 9.6, except that in order to use the prebuilt templates, you need to select the **Display list for Examples** radio button.

Display 9.21 Creating the JDBC TableView Servlet

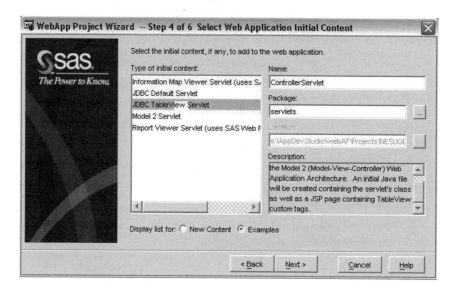

The JDBC TableView Servlet Template creates a complete Web application using the Model 2 (Model-View-Controller) Web Application Architecture. *Models* (including *data vewers*) are objects that provide a way to access data, while *views* provide a visual representation of this data.[6] Models negotiate data transfer via *interfaces*. A local model contains its own data (such as a list of strings), while a remote model retrieves its data from a remote server (such as the SAS AppDev Studio Version 2 DataSetInterface). See http://support.sas.com/rnd/appdev/V2/webAF/ModelView.htm for a detailed explanation of this concept.

In J2EE, a Web application can be divided into three layers—model, view and controller—each with their own responsibilities:

- A model represents business data and business logic or operations that govern access and modification of this business data. Often the model serves as a software approximation to real-world functionality. The model notifies views when it changes and enables the view to query the model about its state. It also enables the controller to access application functionality encapsulated by the model.

- A view renders the contents of a model. It accesses data from the model and specifies how that data should be presented. It updates data presentation when the model changes. A view also forwards user input to a controller.

- A controller defines application behavior. It dispatches user requests and selects views for presentation. It interprets user inputs and maps them into actions to be performed by the model. In a stand-alone GUI client, user inputs include button clicks and menu selections. In a Web application, they are HTTP `get` and `post` requests to the Web tier. A controller selects the next view to display based on the user interactions and the outcome of the model operations.[7]

---

[6] Do not confuse "view" in this sense with the term as it is used in the DATA step and in PROC ACCESS. In this case, we are talking about components that are used to create a user interface. Database views, in contrast, are SQL-based structures that provide user-specific access to data.

[7] Inderjeet Singh, Beth Stearns, Mark Johnson, and the Enterprise Team, *Designing Enterprise Applications with the J2EETM Platform*, 2nd ed. (Boston, MA: Addison-Wesley Professional, 2002).

In this case the *model* is the underlying SAS data, while the *view* is the JSP. The name of the new *controller* servlet logically enough defaults to ControllerServlet.

To create a servlet, follow these steps:

1. Select **JDBC TableView Servlet**, and then click **Next**.

Display 9.22 Selecting the JDBC TableView Servlet Template

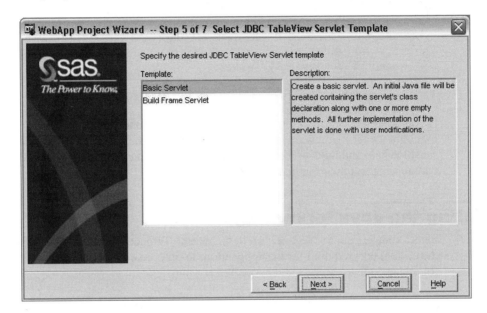

2. Select **Basic Servlet** in Step 5 as shown, then click **Next**.
3. Click **Next** in Step 6, **Specify Basic Servlet Options**.
4. Click **Next** in Step 7, **Specify WebApp Project Options**.
5. Click **Finish** in Step 8 to build the application. This concludes the WebApp Project Wizard.

Modifying the Controller Servlet

The SAS template generates a JavaServer Page **index.jsp** as the view and a controller servlet called **ControllerServlet.java** that uses a JDBC connection to access SAS data on the server. To view the generated servlet, click the **Files** tab in the webAF Project Navigator and then select

```
<webappbase>/WEB-INF\classes\servlets\ControllerServlet.java.
```

You need to make three changes to the generated code in order to point to a specific SAS data set and to allow table editing:

1. Locate the static variable JDBC_DATABASE_URL. Change the value to point to your SAS Workspace Server:

```
private static final String
    JDBC_DATABASE_URL = "jdbc:sasiom://HUNDING:8591";
```

2. Change the value of the variable `jdbcQuery` to read the desired data—in this example the sample shoe sales data set:

```
//Setup the query for the connection,
//  such as "select * from sashelp.class"
String jdbcQuery = "select * from sashelp.shoes";
```

3. Locate the initializer of the variable adapter. Add the following line to allow editing of the data:

```
//Create the model adapter and set it on the session
adapter = new JDBCToTableModelAdapter(
    sas_JDBCConnection, jdbcQuery);
adapter.setReadOnly(false);
```

This completes the necessary changes to the Java servlet code!

If you need more information on any of the Java components in the servlet or JSP, highlight the term you want to search and press F1 for help. SAS has provided context-sensitive help and links to the API to make it somewhat easier to learn about the components and their methods.

Creating the JavaServer Page

Select `<webappbase>/index.jsp` in the webAF Project Navigator. This page contains the custom tags required for the application. In order to enable row-level editing, add the following code to the `TableViewComposite` tag:

```
<sas:TableView>
    <sas:Edit enabled="true" singleRowEditing="false"/>
</sas:TableView>
```

Testing the Application

To build the project using the Java Ant tool,[8] select **Build ▶ Build Project** from the webAF menu, or just press F7.

The sample templates all assume you are using a local SAS server. If your data resides on a remote host, you need to change the properties of the project by selecting **File ▶ Project Properties**. Select **Startup** in the left-hand navigator widow. In the box labeled **For this project, pass additional arguments to the Java interpreter**, change the name of the local host to that of the remote server.

Start the built-in Tomcat 4 servlet engine by going to **Tools ▶ Services ▶ Start Java Web Server**. Now run the application by selecting **Build ▶ Execute in browser** or by clicking on the exclamation point on the Build toolbar.

Depending on the speed of your processor, it can take a very long time to display the page for the first time. This is because the JSP must be parsed, validated, translated into a servlet and finally executed. After the first time the page is displayed, however, it should load much more quickly.

---

[8] Ant is a Java-based build tool for compiling complex applications. For more information about Ant, see the Apache documentation at http://ant.apache.org/manual/.

If you have followed all of these directions, the Web page shown in Display 9.23 should appear in your browser window. Note that the URL for this page points to the **ControllerServlet**, not to **index.jsp**. That is because the servlet is delegating the HTTP request to the page.

Display 9.23 Table of SASHELP.SHOES

The format of this page is determined by a CSS file supplied as part of the default application. In this case, the file is located under the project directory in **webapp\styles\sasComponents.css**. You can change the appearance of the page by editing this style sheet; scroll down about 900 lines to the following comment:

```
TABLEVIEW STYLES (applies to
com.sas.servlet.tbeans.tableview.html.TableView)
```

Changing the font, color or weight of the visual components is just a matter of locating the correct element tag and making the desired changes.

SAS has also provided a template for using what they call *Renderers* to display rows and columns in alternating colors and fonts (see http://support.sas.com/rnd/appdev/examples/ ServletJSP/DisplayingaTableusingJDBC_BLD.htm). This template can be loaded in the same way as the TableView example we have been using.

The next display shows the bottom of the screen. Note the navigation arrows, these are used to scroll up and down over the records in the data set. The three icons in the lower left (✓, **X** and +) run JavaScripts that are generated by the wizard. If the check box in the first column of the page is checked, clicking the ✓ symbol at the bottom of the form deletes the row; otherwise, this action commits any changes made to the record. The **X** symbol cancels any edits, while the + symbol is used to add a new record to the data set. Clicking on a column heading brings up a menu for sorting the data or moving the column. Scroll up to the top of the page, as shown in Display 9.23. The small symbol in the upper left can be used to export the entire data set to Microsoft Excel.

Display 9.24 Table of SASHELP.SHOES (*continued*)

Deploying the Web Application

Sending the page to another Web server is made simple by the webAF Package Wizard. Note that the only version of Tomcat formally supported by SAS is Version 4.18; anything newer (Tomcat 5, for instance) is not guaranteed to work. The example below was run in Tomcat 4.30.

Note also that the application has to have access to the SAS server. For that reason, the connection to the local host should use the SAS server name, not a local address like 127.0.0.1.

To create a Web archive (WAR) file from the application in webAF, just click on **Tools ▶ Wizards ▶ Package Wizard**. There are only two steps; you probably want to accept the defaults, so just click **Next** each time. The WAR file should contain all of the components needed to deploy the file.

Now open the Tomcat Manager application on the Web server. Go to **Upload a WAR file to install**. You should be able to browse to a file called **Demo.war** in the project directory. Click **Install**. The WAR file (which can be quite large) will be copied to the server and deployed. If all goes well, you should get the following message:

```
OK - Installed application at context path /Demo.
```

Now enter the URL for the servlet, in this case the following:

```
http://<hostname>:8080/Demo/ControllerServlet
```

Again, it may take some time for the contents of the WAR file to be extracted. You should see the same page as shown in Display 9.23. Congratulations. You have now created and deployed a Java Web application.

Using SAS Remote Compute Services

In addition to accessing SAS data, it is also possible to run SAS programs remotely. The SAS AppDev Studio 3.1 tag library is an addition to, rather than a replacement for, the older component palette. Chapter 11, "Building Web Applications with SAS and Java," explains how to use stored processes to run SAS jobs on a remote host. This chapter covers some of the older but still useful interfaces that were part of SAS AppDev Studio 2.0, in particular the SAS SubmitInterface and the DataSetInfo interfaces.

Remote Computing Using SubmitInterface

The SubmitInterface control can be used to send SAS program statements to the server. The following program revisits the retail data set used in the CGI examples. The SAS source code for this program is the same as that used in Chapter 6, "SAS/IntrNet: the Application Dispatcher."

Example 9.5 Sample PROC REPORT Macro

```
options nodate nonumber noovp nocenter pagesize=20;

/* Sample Program: shoes.sas      */

%macro salesrpt(region);

   proc report data=sashelp.shoes;
      by region;
      %if ( &region ne null ) %then %do;
            where region="&region";
      %end;
      title "<h2>Shoe Sales by Region x Product</h2>";
      footnote "Data are current as of &systime &sysdate9";
      column product sales;
      define product / group;
      define sales / analysis sum;
   quit;

%mend salesrpt;

%salesrpt(<%= request.getParameter("region") %>)
```

The code has been modified from that shown earlier in that the macro variable REGION is read from the URL as a parameter. (The macro code is shown in bold in the example.) If the parameter is not supplied, the default is to run the report for all of the observations in the data set. This example illustrates that it is actually possible to insert Java code within the SAS program.

The following JSP code can be used to send this program to the SAS server:

Example 9.6 JavaServer Page Using SubmitInterface

```
<%@ taglib uri="http://www.sas.com/taglib/sasads"
      prefix="sasads" %>
<%@ include file="header.html" %>
<body>
   <h1>SubmitInterface Example</h1>
   <sasads:Connection
        id="connection1"
        serverArchitecture="IOM"
        host="hunding"
        port="8591"
        username="sas"
        password="sasuser">
      <sasads:Submit
           connection="connection1"
           display="LASTOUTPUT" scope="page">
          <%@ include file="shoes.sas" %>
      </sasads:Submit>
   </sasads:Connection>
</body>
<html>
```

In order to use the SAS AppDev Studio 2.0 `sasads:Connection` and `sasads:Submit` custom tags, the taglib directive for this library must be included in the page. When creating the project, specify the **SAS Taglib** and **Tbeans (Version 2)** run-time classes in the New Project Wizard. This will add a JSP `taglib` directive as shown in Example 9.7:

```
<%@ taglib uri="http://www.sas.com/taglib/sasads"
prefix="sasads" %>
```

To build the JSP, first add the connection control shown, with the following attributes:

- `id` – a unique name for the connection
- `serverArchitecture` – the connection type, in this case `IOM`
- `host` – the name of the remote host
- `port` – the port on which the spawner is listening; 8591 is the default for a workspace server
- `username` – a valid user on the host system
- `password` – a valid password, must be included in the JSP

For this example, the connection `scope` parameter defaults to `page`, so the connection is closed with the HTML page; specifying `scope="session"` maintains the connection for the duration of the HTTP session.

The SubmitInterface component takes the following attributes (the defaults are shown in bold):

Table 9.2 SubmitInterface Attributes

Attribute Name	Value	Comment
connection	"java.lang.String"	Valid connection to a SAS/CONNECT server
connectionObject	"com.sas.rmi.Connection"	*InformationBean*: reference a subclass if necessary
display	"NONE|LOG|LASTLOG|OUTPUT|LASTOUTPUT"	Specify output type
format	"MONOSPACE|SAS"	Default value inserts <pre> </pre> tags around the output; otherwise HTML formatting is lost
id	"value"	Case-sensitive name used to identify the object instance
ref	"java.lang.String"	Object created earlier in the same scope; alternative to **id**
scope	"page|request|session|application"	The scope (or page context) within which the reference is available

In the preceding example, the display format is specified as LASTOUTPUT; in this way only the current report is shown. Otherwise resubmitting the page would result in cumulative multiple-output reports.

Display 9.25 PROC REPORT Output

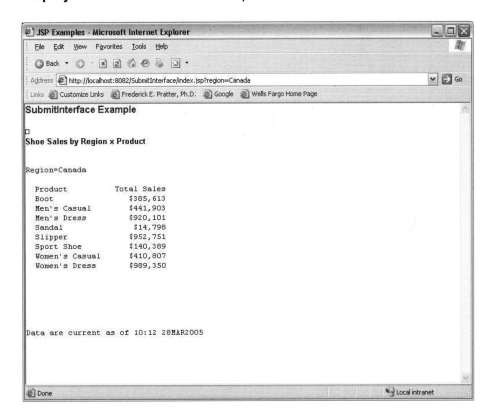

The URL of this JavaServer Page is

```
http://localhost:8082/SubmitInterface/index.jsp?region=Canada
```

The server name `localhost:8082` refers to the development Java Web server started from webAF software. Note that the data is being served up dynamically by the remote host; these values will always be the most current, since the REPORT procedure is run by the SAS server.

The `region` value is supplied as a parameter, which in turn is passed to the SAS program as a macro variable value. (Notice that it is not necessary to enclose the parameter value in quotation marks.) Specifying a different value for the `region` parameter displays the report for that region.

The plain appearance of the text is a result of using the default monospace format. There is no simple way to get a customized look; because of the preformatting, CSS stylesheet values are ignored. In order to use ODS with the SubmitInterface component, the application has to use the socket access method to communicate with the server. The output from ODS can then be streamed from SAS to the server. The details of doing this are more complex than is appropriate for discussion here; see the example program **streamingODSOutput.jsp** at http://support.sas.com/rnd/appdev/V2/webAF/ server/examples/streaming2.htm for more information.

Using the DataSetInfo Interface

The program shown in Example 9.6 has one major flaw, which is that it is necessary to know in advance the set of possible values for the REGION variable (and the correct case and spelling of the values). Instead, it is possible to create a model that contains an up-to-date list of regions by using the DataSetInfo interface in SAS AppDev Studio 2.0. The following program sets up a combo box that allows the user to select one of the region codes from the shoe sales data.

Example 9.7 Using the SAS AppDev Studio 2.0 DataSetInfo Interface

```
<%@ taglib uri="http://www.sas.com/taglib/sasads"
    prefix="sasads"%>
<%@ page import=
    "com.sas.sasserver.datasetinfo.DataSetInfoInterface" %>
<%@ page import="com.sas.collection.OrderedCollection" %>
<%@ include file="header.html" %>

<sasads:Connection
  id="connection1"
  serverArchitecture="IOM"
  scope="session"
  host="hunding"
  port="8591"
  username="sasadm"
  password="system"/>
```

```
<% // Java scriptlet to add choicebox values to page context
   DataSetInfoInterface dsinfo = (DataSetInfoInterface)
        com.sas.servlet.util.Util.newInstance
            (connection1.getClassFactory(),
             connection1,
             DataSetInfoInterface.class);
   dsinfo.setDataSet("SASHELP.SHOES");

   // display unique values of region
   int index = dsinfo.getVariableIndex("REGION");
   pageContext.setAttribute("values",
       new OrderedCollection
           (dsinfo.getVariableUniqueValues(index)));
%>

<div align="center">

<h1>International Shoe Sales Data</h1>

<sasads:Form action="index.jsp" >
  <sasads:Choicebox
        id="region"
        model="values"
        prolog="<strong>Select region for report: </strong>" />
  <sasads:PushButton id="submit" text="Submit" />
</sasads:Form>

</div>
```

This program illustrates the use of the `DataSetInfo` interface to return information about the remote data set. This component is not available from the Component Palette; it is necessary to import the class with a page directive as illustrated. The required Java scriplet code is shown in bold.

The `newInstance` method of the class `com.sas.servlet.util.Util` creates an instance of a remote model using the class factory and connection given; a Java object is returned. The three arguments are:

- rocf (Remote Object Class Factory) – `connection1.getClassFactory()`
- connection – `connection1`
- class – a new instance of `DataSetInfoInterface.class`

The packages available in webAF software are listed in the **All Classes** tab of the Project Navigator. Scrolling down to the utility package shows that the following interfaces and classes are available:

Display 9.26 webAF AllClasses Tab

Right-clicking the **Util** tab and selecting **Help** displays the Java API for the class, as shown in Display 9.27.

Display 9.27 Java API for `com.sas.servlet.util.Util`

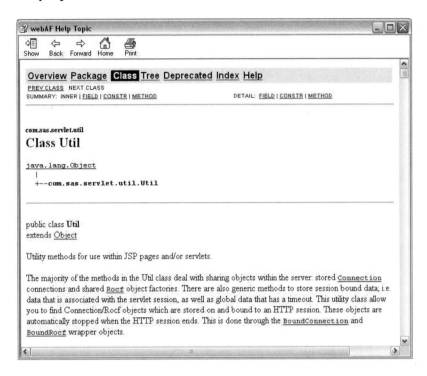

This raises the question of how you know that, in order to create a new DataSetInfoInterface instance, it is necessary to cast an object created with the `newInstance()` method of this utility class. The answer is that you just have to know that the method returns a generic object:

```
public static Object newInstance
   (Rocf rocf, Connection connection, Class c)
```

As with most Java programming, the way to learn what classes and methods are available and how to use them is to look at a lot of examples in the documentation (and of course read this book).

In any case, having instantiated the `DataSetInfoInterface`, it is now possible to take advantage of the methods provided to get the unique values of the REGION variable.

- `dsinfo.setDataSet("SASHELP.SHOES")` – opens the SHOES data set
- `int index = dsinfo.getVariableIndex("REGION")` – specifies the column number of the REGION variable
- `dsinfo.getVariableUniqueValues(index)` – extracts unique values of REGION into a String array

The array is then assigned to the page context as an array of values, which in turn are used as the model for the list box, as shown in the preceding section. The resulting page is shown in Display 9.28.

Display 9.28 Data-Driven Combo Box

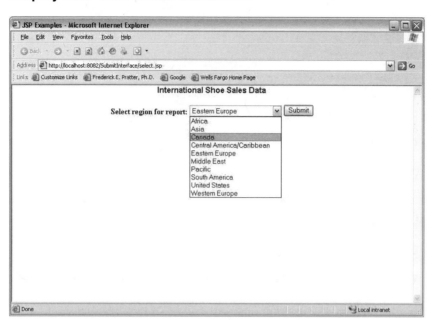

Since the value of the list box `id` is `"region"`, clicking on the **Submit** button sends the selected text value to **index.jsp** as a parameter, resulting in the output shown in Display 9.28. Incidentally, since the `Connection` object is created in this page with session scope, it is not necessary to include this tag in **index.jsp**. When subsequent pages are referenced in the same session, the connection to the remote server will be available.

The two components used in this example—TableView and DataSetInterface—are only two of the choices from the webAF menu. Each of these components is a Java class in one of the `com.sas` packages. SAS has provided class documentation for the SAS AppDev Studio 2.0 classes at http://support.sas.com/rnd/appdev/V2/webAF/api/, but unfortunately there is no available description of what they do, except as embedded in the Java documentation.

Model components are accessed from the **SAS** tab on the Component Palette and include the following set of interfaces:

- CatalogEntryListInterface – retrieves a list of SAS catalog entries, or SAS catalog members, from a SAS session

- CatalogListInterface – retrieves a list of SAS catalogs from a SAS session

- DataSetListInterface – retrieves a list of SAS data sets from a SAS session

- DataSetV2Interface – defines an interface to a SAS data set (extends the DatasetInterface while adding additional functionality)

- FormatInterface – defines an interface for formatting a set of values

- LevelTreeInterface – talks to the SAS server to create a LevelTree which can be used to display a SAS level data set

- LibraryListInterface – retrieves a list of SAS libraries from a SAS session

- MdTo2dTable – models a multidimensional table as a two-dimensional table

- MultidimensionalTableV3Interface – defines an interface to SAS multidimensional data

- ParentChildInterface – defines an interface to the Parent/Child hierarchical model
- SASFileListInterface – retrieves a list of SAS library members, referred to as SAS files, from a SAS session
- SASProcedureTemplate – creates SAS PROC statements
- SchemaListInterface – retrieves a list of schemas in a database
- SclFuncsV3Interface – defines an interface to SAS Component Language (SCL) functions
- SubmitInterface – sends and retrieves SAS statements to and from the SAS Program Editor, Log, or Output windows
- SummaryInterface – defines an interface to the SAS data set summary model
- SystemNodeInterface – no explanation of use
- TableListInterface – retrieves a list of tables in a database catalog or schema

The views available on the remaining tabs include a mixture of Swing components and components supplied by SAS:

- Selector – `JButton`, `JCheckBox`, `JComboBox`, `JList`, `JProgressBar`, `JScrollBar`, `JSlider`, `JToggleButton`, `JToolbar`, `JTree`, `JRadioButton`, `ButtonGroup`, `TriStateCheckBox`, `CheckBoxList`, `DualListSelector`, `DualTreeSelector`, `TreeListSelector`
- Text – `JLabel`, `JTextField`, `JTextArea`, `JTextPane`, `JPasswordField`, `JSpinner`
- Data Viewers – `JTable`, `TableView`, `NavigationBar`, `SortableTableModelAdapter`
- Graphics – `PieChart`, `BarChart`, `BarLineChart`, `LineChart`, `LinePlot`, `Scatterplot`, `RadarChart`, `ImageView`, `RangeView`
- Container – `Jpanel`, `JScrollPane`, `JSplitPane`, `JTabbedPane`
- Utility – `JColorChooser`, `JFileChooser`

As the preceding example illustrates, first select a model and associate it with the frame and with the default page remote connection, then add view components to the form and link them to the interface by using the Customizers for each control. SAS has implemented an object-oriented programming interface using custom components to make data access straightforward if not easy.

Using the DataBean Wizard

In addition to the approach outlined in the preceding sections, webAF software also offers an alternative methodology for data access. Using the *DataBean* Wizard, you can create a custom interface with the properties desired for a particular application. The following example creates a new JavaServer Page that uses a Bean for data.

First, open webAF software and create a new Web Application project as in the preceding example. Then select **Tools ▶ Wizards ▶ DataBean Wizard** from the main menu. (Unless you have a project open you cannot access this wizard.)

The first setup screen displays the following message:

> The DataBean wizard will assist you in creating a Java data class that will provide access to a specific data table. In the following steps you will be asked to provide information to identify the location of the data table and what attributes of that table should be included in the Java data class.

The screen offers a choice of database connections: **JDBC** or **SAS DatasetInterface**. This example illustrates how to access a SAS data set; with a JDBC connection you can easily build a DataBean that can support access to another RDBMS (such as Oracle, Sybase or SQL Server). You do not need to have the SAS/ACCESS engine installed to use JDBC.

The DataBean Wizard goes through the following seven steps:

1. In the first screen, choose **SAS Dataset via DataSetInterface**.
2. Select the remote connection from the list of registered connections.
3. Select the **SASHELP** library and the **RETAIL** data set.
4. The **Advanced** button allows you to specify DATAFORM entry and source; leave this box unchecked unless you have an SCL Data Form you want to use.
5. Select the columns from the table. The default is to use all of them, but you can deselect some if you are not going to need them here.
6. Specify the name for the new class. The default is `library-name_data set-name`—in this case, `Sashelp_retail`.
7. The final screen summarizes the instructions for creating the Bean, as shown in Display 9.29. Click **Finish** to accept, or **Back** to revise.

Display 9.29 Adding a DataBean to the Project

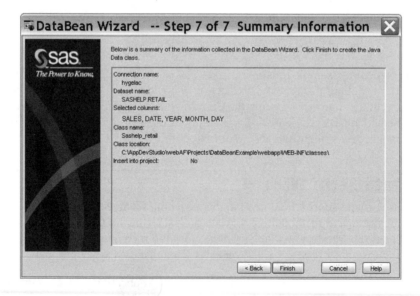

The SAS AppDev Studio documentation at http://support.sas.com/rnd/appdev/examples/ServletJSP/DataBeanForm_abt.htm provides detailed instructions on how to build a Web Application using the resulting DataBean. Click **Build It** to see how to construct a JavaServer Page using the components provided by SAS.

Conclusion

Up to this point, all of the code examples shown work equally well in SAS 8 and SAS®9. The following section describes the new features available in SAS Integration Technologies for SAS®9. Chapter 10, "Using the SAS Open Metadata Architecture with the Integrated Object Model," introduces the SAS Metadata Server, while Chapter 11, "Building Web Applications with SAS and Java," illustrates how to use stored processes as a replacement for SAS/IntrNet as well as the SAS AppDev Studio 2.0 interfaces described previously. The remaining chapters then briefly introduce some of the additional features of SAS Integrations Technologies including SAS BI Web Services and Portals. These new components are considerably more complex than those described thus far, and require advanced Java programming skills. For Java programmers who want to use SAS data and remote computing facilities, SAS Integration Technologies provides a flexible and powerful set of interfaces.

References

Java Documentation

URL references are current as of the date of publication.

- Singh, Inderjeet, Beth Stearns, Mark Johnson, and the Enterprise Team. 2002. *Designing Enterprise Applications with the J2EETM Platform.* 2nd ed. Santa Clara, CA: Sun Microsystems, Inc. http://java.sun.com/blueprints/guidelines/designing_enterprise_applications_2e

- Sun Microsystems, Inc. 2005. *The Java Web Services Tutorial: For Java Web Services Developer's Pack, v1.6.* Santa Clara, CA: Sun Microsystems, Inc. http://java.sun.com/webservices/docs/1.6/tutorial/doc/

SAS Documentation

- Basile, Aimee, and Dave Hayden. 2004. "Architecting AppDev Studio-Based Web Applications with Object Oriented Methodologies." *Proceedings of the Twenty-Ninth Annual SAS Users Group International Conference.* Cary, NC: SAS Institute Inc.

- Benson, Corey, and Robert Girardin. 2004. "A Guide to Understanding Web Application Development." *Proceedings of the Twenty-Ninth Annual SAS Users Group International Conference.* Cary, NC: SAS Institute Inc.

- Boudreaux, Don. 2004. "Java Syntax for SAS Programmers." *Proceedings of the Twenty-Ninth Annual SAS Users Group International Conference.* Cary, NC: SAS Institute Inc.

- Cisternas, Miriam G., and Ricardo A. Cisternas. 2004. "Java Servlets and Java Server Pages for SAS Programmers: An Introduction." *Proceedings of the Twenty-Ninth Annual SAS Users Group International Conference.* Cary, NC: SAS Institute Inc.

- Curnutt, Randy, Michael J. Pell, and John M. LaBore. 2002. "Energizing End Users with a Slice of SAS and a Cup of Java." *Proceedings of the Twenty-Seventh Annual SAS Users Group International Conference.* Cary, NC: SAS Institute Inc.

- DeMartino, Heather E. 2002. "Keeping Form Data from Falling into the Bit Bucket with webAF." *Proceedings of the Twenty-Seventh Annual SAS Users Group International Conference.* Cary, NC: SAS Institute Inc.

- Ferguson, Chad, and Sandra Brey. 2004. "Developing Data-Driven Applications Using JDBC and Java Servlet/JSP Technologies." *Proceedings of the Twenty-Ninth Annual SAS Users Group International Conference.* Cary, NC: SAS Institute Inc.

- Girardin, Robert. 2003. "Introduction to the SAS Custom Tag Library." *Proceedings of the Twenty-Eighth Annual SAS Users Group International Conference.* Cary, NC: SAS Institute Inc.

- Herbert, Pat. 2001. "Delivering Information Everywhere Using JSP and SAS." *Proceedings of the Twenty-Sixth Annual SAS Users Group International Conference.* Cary, NC: SAS Institute Inc.

- Hoyle, Larry, and Mickey Waxman. 1999. "SAS webAF for Java Application Development, a First Sip." *Proceedings of the Twenty-Fourth Annual SAS Users Group International Conference.* Cary, NC: SAS Institute Inc.

- Main, Rich. 2004. "Exploiting the SAS Business Intelligence Architecture Using SAS AppDev Studio 3. 0." *Proceedings of the Twenty-Ninth Annual SAS Users Group International Conference.* Cary, NC: SAS Institute Inc.

- Main, Rich. 2002. "Introduction to AppDev Studio Version 3.1." *Proceedings of the Twenty-Seventh Annual SAS Users Group International Conference.* Cary, NC: SAS Institute Inc.

- Nicklin, Clare A., and Daniel Morris. 2004. "A Successful Implementation of a Complicated Web-Based Application through webAF and SAS Integration Technologies." *Proceedings of the Twenty-Ninth Annual SAS Users Group International Conference.* Cary, NC: SAS Institute Inc.

- Pratter, Frederick. 2004. "Building an Online Entry Form with WebAF (and a Little Java)." *Proceedings of the Twenty-Ninth Annual SAS Users Group International Conference.* Cary, NC: SAS Institute Inc.

- SAS Institute Inc. 2003. *SAS AppDev Studio 3.0 Migration Guide.* Cary, NC: SAS Institute Inc.

- Sipe, Lori L., and Qing Chen. 2002. "Creating a Web-Based Application Utilizing JSP and SAS JAVA BEAN." *Proceedings of the Twenty-Seventh Annual SAS Users Group International Conference.* Cary, NC: SAS Institute Inc.

- Stevens, Jay L., and Brian Santucci. 2001. "Integrating SAS with an Open World: Java, JSP, LDAP, and Oracle." *Proceedings of the Twenty-Sixth Annual SAS Users Group International Conference.* Cary, NC: SAS Institute Inc.

Part 4

SAS Integration Technologies

Chapter 10

Using the SAS Open Metadata Architecture with the Integrated Object Model

Overview: SAS Integration Technologies

SAS Integration Technologies provides access to SAS data and procedures from a variety of clients, including Java, Microsoft Office, Visual Studio, ASP, and the Web, as well as from SAS components. Prior to SAS 8, it was possible to access SAS from other applications in a limited manner with DDE or OLE automation. SAS Integration Technologies adds a set of powerful and complex tools allowing developers to interact with SAS using the *Integrated Object Model* (IOM) interface. The IOM can be used to connect to SAS from *open clients*, which are nonproprietary solutions using standard programming languages such as Java, Visual Basic or C++.

SAS defines the Integrated Object Model as a set of "distributed object interfaces to SAS.... IOM enables you to use industry-standard languages, programming tools, and communication protocols to develop client programs that access these services on IOM servers." (*SAS Integration Technologies: Technical Overview*). This chapter is intended to introduce SAS Integration Technologies (and the IOM in particular) to that segment of the community of SAS users for whom the very concept of a distributed object interface is unfamiliar and potentially scary.

The success of SAS over the last 30 years in no small part has been because very powerful results can be obtained with a relatively small investment in programming effort. The IOM comes at the problem from the other side: if you already know how to program in, say, Visual C++, the IOM can allow you to access SAS data and run SAS procedures from within a C++ program. This is an extremely important addition to the overall functionality of SAS, but the range of choices available and the complexity is such that many users may be dissuaded from attempting to enter into this new realm. SAS has provided a number of roadmaps for SAS Integration Technologies (see the references at the end of this paper), but a roadmap is of use only if you know where you want to go. This chapter and the one that follows attempt to begin a little further back, by suggesting a number of interesting and rewarding destinations.

SAS Integration Technologies includes a number of integration and development tools in addition to the IOM. SAS 8 introduced several kinds of IOM servers, the Publishing Framework (PUSH capability), the Application Messaging interface, and the Directory Service interface for access to LDAP services. SAS®9 added SAS Management Console along with production versions of SAS Foundation Services (for Java programming), the SAS Stored Processes application interface, the SAS Web Infrastructure Kit for building portals, and SAS BI Web Services. All of these are designed to make use of the new SAS Open Metadata Interface. (See "SAS Integration Technologies: What's New?" at http://support.sas.com/rnd/itech/updates/new.html.)

In this chapter, the Integrated Object Model (IOM) is described in the context of the SAS Open Metadata Architecture (OMA). Chapter 11, "Building Web Applications with SAS and Java," covers Java Web applications, including the SAS Stored Process Web application. All of these are based on the new SAS Open Metadata Architecture and require an understanding of the IOM, so that is where we shall begin.

The Integrated Object Model

The first thing to understand about the Integrated Object Model is that it is designed to work in a client-server environment, that is, one in which each computer on the network performs some specific function, such as managing data or Web page requests. In this model, the user interface, functional process logic and data access are each handled by separate modules, most often on separate computers. This separation between the presentation layer, business rules, and implementation is central to modern reusable software design.

The following diagram, from the SAS 8 Enterprise Integration documentation, illustrates one such scenario (it gets reproduced a lot):

Display 10.1 The Integrated Object Model

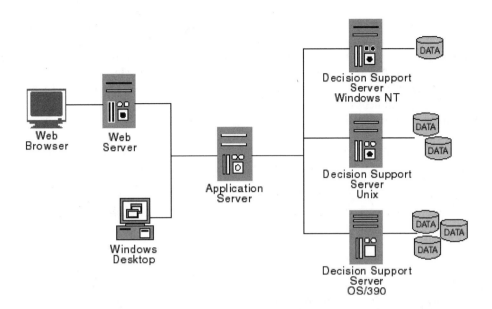

In the traditional model of SAS computing, programs that access and manipulate data all ran on the same computer as the data, whether on a PC, a UNIX machine, or a remote mainframe over a terminal connection. SAS/SHARE software and to some extent SAS/CONNECT software made it possible to connect clients to SAS servers over a network, but as the preceding diagram suggests, SAS Integration Technologies can be used to automate and distribute SAS data processing tasks over a wide variety of platforms.

One important point, which the SAS documentation assumes but does not ever say explicitly, is that the IOM is primarily designed for thin client applications. If all your users have SAS on their local desktops, you do not need the IOM. The exception to this is SAS Enterprise Guide software, which communicates with SAS using the IOM when SAS and SAS Enterprise Guide run on the same machine, but in a thin client world. Users who need access to SAS data or procedures can obtain it via the Web or through a user interface constructed with any of the common development tools such as Visual Basic, C++, and even Microsoft Office.

A second point that should be emphasized is that the IOM provides remote users with access not only to SAS data but also to the full capabilities of the procedures in the Base SAS language. Most relational database management systems, such as Oracle or SQL Server, have a scripting language for data manipulation in addition to retrieval, but none of these has the power of SAS to carry out complex analyses. The Oracle PL/SQL language, for example, although quite powerful in its own right, is hardly capable of running the GLM procedure or the NLIN procedure!

So in general, when do you need SAS Integration Technologies? If you are distributing thin client applications that require complex statistical and graphical capabilities, then the SAS Integrated Object Model provides an elegant solution. If you just want to be able to copy the data from SAS tables into Microsoft Excel, you probably do not need to use the IOM to do it (although you can).[1]

---

[1] See Peter Eberhardt, "Rev Up Your Spreadsheets with Some V8 Power," *Proceedings of the Twenty-Seventh Annual SAS User's Group International Conference* (Cary, NC: SAS Institute Inc., 2002).

IOM Servers

Another important difference between the client-server world and the traditional SAS model is that SAS Integration Technologies assumes that you have two kinds of people on your SAS programming staff: developers and system administrators. This is reflected in the materials provided by SAS. As the references at the end of the chapter indicate, there are currently eight separate manuals, provided for different classes of users (see "SAS Integration Technologies: Software and Documentation Downloads" at http://support.sas.com/rnd/itech/updates/index.html for the most up-to-date editions):

- *SAS Integration Technologies: Technical Overview*
- *SAS Integration Technologies: Server Administrator's Guide*
- *SAS Integration Technologies: Administrator's Guide*
- *SAS Integration Technologies: Administrator's Guide (LDAP Version)*
- *SAS Integration Technologies: Developer's Guide*
- *SAS Web Infrastructure Kit: Overview*
- *SAS Web Infrastructure Kit: Administrator's Guide*
- *SAS Web Infrastructure Kit: Developer's Guide*

Finding a specific solution sometimes requires a lot of patience as you navigate a web of interrelated topics. Alas, as is all too frequently the case with system documentation (and SAS is by no means the worst offender) the documentation assumes that you already understand what all this stuff does. Setting up the necessary SAS servers is not a task for the faint of heart. It requires a substantial amount of experience both with SAS and with the specific platform employed, whether Windows, UNIX or mainframe. The documentation is written for programmers, not for users, and is necessarily complex given the range of possible options available and the lengthy list of supported features.

Consequently, the initial installation for SAS Integration Technologies should be done by your SAS system administrator (if you are lucky enough to have one). Do not attempt to do this process on your desktop system; designate one or more separate SAS servers for your network, and set aside a block of time for completing all the configuration tasks. It's not hard, it just takes time. Fortunately, you only have to do it once. Incidentally, this is where all those extra SAS user IDs (*sasadm, sassrv, sasguest, sastrust, saswdadm* and *sasdemo*) recommended in the installation documentation actually become necessary.

Once the SAS server-side components have been installed, the system administrator will need to run the Configuration and Management Wizard, included on a separate disk in the installation materials, to define all of the different kinds of SAS Integration Technology servers. (You can configure a system without the wizard, but it is a lot easier if you use it; see "Getting Started Without the SAS Configuration Wizard" in the *SAS Integration Technologies: Server Administrator's Guide* at http://support.sas.com/rnd/itech/library/toc_adminoma.html.

The configuration wizard generates a lengthy HTML document called `My Configuration Steps.htm`. This document details the numerous specific steps required to complete the deployment. The first step is to start the metadata server. On Windows, the SAS Metadata Server should be configured to run as a service that is started automatically at boot time. On UNIX or mainframe systems, the system administrator can set up the server to start automatically using the appropriate procedures for the operating platform. Consequently, you probably will not need to manually start or stop the metadata server, or even worry very much about it, after this first time.

The subsequent steps all use the new SAS Management Console (available on the client-side installation disks) to set up metadata repositories, users, authorizations, libraries and servers. As the SAS documentation points out:

> SAS Multi-Vendor architecture lets SAS run on virtually any platform. The SAS Management Console is Java-based and will run on Windows and UNIX platforms. It doesn't have to run on the system where the application or SAS application server is running. It can be run from an administrator's desktop system, yet manage the resources on all SAS supported platforms.

You can also use the (Windows-only) SAS Enterprise Guide Administrator to manage server connections. Note that as the FAQ indicates, you do not need to have SAS installed on the client system to use these application managers. SAS Management Console and SAS Enterprise Guide work perfectly well connecting to SAS on a remote host.

When you start SAS Management Console, you will need to specify a metadata profile that includes the server to which the definitions will be written, the active metadata repository, and any required connection information such as user IDs and passwords. At the conclusion of the approximately 28-step process detailed in the configuration file, SAS Management Console should look a lot like Display 10.2 (depending on which products you have licensed and installed):

Display 10.2 SAS Management Console

In order to make sense of all of the options provided by SAS Management Console, it is important to understand that the Integrated Object Model is implemented on the client. In other words, as a developer you need to write a program (in SAS, Java, C++ or some other general purpose programming language) that instantiates this interface and that runs on an end user's PC.

However, to make use of the IOM, it is necessary to have one or more object servers running on the network.

Four kinds of IOM servers are provided, all of which can be managed with SAS Management Console:

- The SAS Metadata Server stores and manages metadata repositories, which store data about the SAS servers, libraries and stored processes available.
- The SAS Workspace Server supports the SAS programming environment, and is functionally equivalent to a Display Manager session.
- The SAS Stored Process Server runs "canned" SAS programs.
- The SAS OLAP Server delivers data cubes (presummarized tabulations) to SAS Enterprise Guide or other OLAP clients.

The SAS OLAP Server is a specialized platform for data access provided with SAS Intelligent Storage products. The other three kinds of object servers are more general in scope. It is useful for the developer to have a clear idea of the functions and requirement for each.

The SAS Metadata Server

As previously noted, in order for a client program to take advantage of distributed objects, it is necessary to have an instance of SAS running on the host. A SAS object server is an enhanced database engine, essentially SAS/SHARE software "on steroids." The difference is that while SAS/SHARE software supports only remote library services, an IOM server provides the full functionality of a Display Manager session, by exposing SAS objects that provide data and methods to the client.

SAS 9.1 is based on an entirely new model for storing and managing information about available data sources, business rules, and security authorizations. This model is called the SAS Open Metadata Architecture. Within this architecture, the SAS Metadata Server is a shared software application that provides access to metadata that is, to information about data, stored in one or more Metadata Repositories on the server. The SAS Open Metadata Interface is the API for accessing the metadata server from a variety of environments, including Java, Windows applications, and of course from SAS programs.

The SAS Open Metadata Architecture is intended to provide a single, central point of access for all of the information required by an organization, whether a business, a university, or a research group. The advantages claimed for this approach are that it simplifies system support and documentation and helps to ensure data integrity. The disadvantage is that it makes everything more complicated for the SAS system administrator (and to some extent, for the end user). However, once the metadata server is set up and configured, and procedures for accessing it are documented and disseminated, the result should be greater efficiency and lower maintenance costs.

The SAS Workspace Server

As Display 10.2 illustrates, the SAS Metadata Server is used to manage connections to one or more logical workspace servers along with OLAP servers, stored process servers, and SAS/CONNECT and SAS/SHARE software if they are licensed. The SAS Workspace Server is the actual connection to a SAS session. While it is technically possible to start a workspace server by itself (this was the usual way in SAS 8), generally you will require an initial call to a SAS Metadata Server first. This is because in order to use a specified SAS Workspace Server, the developer needs to know how to connect to the host and what information is available there. In the new SAS Open Metadata Architecture, this information is stored in one or more *metadata repositories* and managed by SAS Management Console, as noted previously.

There are three kinds of workspace servers, depending on the client application and the server operating system:

1. Java clients connect to IOM servers using the *IOM Bridge for Java* provided by SAS.

2. Windows client applications can connect to IOM servers running in a Windows operating environment by using the Microsoft *Component Object Model* (COM) as the server protocol. If the object server and the client are on the same system, the connection uses the COM; if the server is on a remote host, the connection is made using the Microsoft *Distributed Component Object Model* (DCOM). (See http://www.microsoft.com/com/ for more information about the range of Microsoft component software models, including COM, COM+, DCOM and ActiveX controls.)

3. There is also an *IOM Bridge for COM* that allows Windows clients to access servers running on UNIX or a mainframe system such as z/OS. It can also be used to make distributed Windows-to-Windows connections, and SAS recommends this approach over the DCOM since it is somewhat easier to administer.

For Windows applications, the IOM supports the OLE DB access protocols that are used by Active Data Objects (ADO). In Java environments, the JDBC 2.0 access protocol is supported. Whatever method is used, the principal interfaces of the SAS Workspace Server include the following:

- Workspace – represents the SAS session
- Language Service – submits SAS DATA and PROC steps, retrieves LOG and LIST output, runs stored processes
- Data Service – manages SAS LIBREFS
- File Service – manages FILEREFS
- Utilities – provide formats, options, result packages, host system information

The section on client implementations illustrates the use of the first two of these: the Workspace and the Language Service. The interested reader is referred to the *SAS Integration Technologies: Developer's Guide* for more information and examples about all of these.

The SAS Stored Process Server

The third kind of IOM server available for applications development is a SAS Stored Process Server. This is a lot like the SAS Workspace Server, with one important difference. SAS Stored Processes are slightly different from traditional SAS programs in that they must be executed on a SAS Integration Technologies server. They cannot be executed in a normal batch or interactive SAS session. (See "SAS Stored Processes – A Roadmap" at http://support.sas.com/rnd/itech/papers/.)

The easiest way to think of stored processes is that they are just "canned" SAS programs that can be run using input parameters supplied at run time in the form of macro variables. The main difference between running a SAS job in a workspace and running it as a stored process is that with the former approach you can generate and submit code from the client *or* run code already existing on the server, whereas the stored process server requires code to be accessible from the server.

Stored process results can be made available in a variety of ways: via the Web (using the SAS Stored Process Web Application, via the SAS Information Delivery Portal or with SAS BI Web Services), in Microsoft Office (with the SAS Add-In for Microsoft Office software), or with SAS Enterprise Guide. It is also possible to access stored processes programmatically on a workspace server, as the example in the following section shows. Chapter 11, "Building Web Applications with SAS and Java," goes into more detail on using the SAS Stored Process Web Application and the SAS Stored Process Service.

IOM Client Applications

Examples of three IOM client applications will be illustrated in this section, the Windows client interface with a Visual Basic example, the Visual C++ interface, and the Java client interface.

Windows Client Interface – Visual Basic Example

The easiest way to get started programming with the Integrated Object Model is probably in Visual Basic. The following example illustrates how to run a SAS program to display the records from a data set in a VB Message Box. Obviously, it is possible to update as well as retrieve information from the server and, with ODS, to return SAS output in a variety of formats. This example demonstrates the basic principles for connecting to a remote host and displaying information.

Display 10.3 Simple Visual Basic Form to Test an IOM Connection

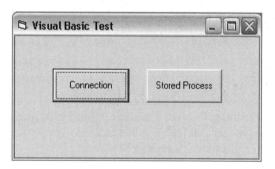

A trivial Visual Basic form was constructed (using Visual Studio 6 SP5) as shown in Display 10.3.

In order to use the IOM from a Visual Basic project, it is first necessary to add references to the SAS link libraries. To add these, go to **Project ▶ References** on the main Visual Basic menu. Scroll down the list and click to add the following references:

- SAS: Integrated Object Model (SAS System 9.1)
- SASWorkspaceManager 1.1 Type Library

This will allow Visual Basic to resolve references to the SAS objects used to connect to the server. The code behind this form is shown in Example 10.1. (This program borrows heavily from the first example in Jahn, 2004.)

Example 10.1 Visual Basic Code to Test a DCOM Workspace Connection

```
Option Explicit
' define a global workspace
Dim obSAS As SAS.Workspace
Dim obWSMgr As New SASWorkspaceManager.WorkspaceManager
Private Sub Form_Load()
     Dim xmlInfo As String
     ' create Workspace server
     Dim obServer As New SASWorkspaceManager.ServerDef
     obServer.MachineDNSName = "hunding"
     Set obSAS = obWSMgr.Workspaces.CreateWorkspaceByServer _
         ("", VisibilityProcess, obServer, "", "", xmlInfo)
End Sub
Private Sub cmdTest1_Click()
     ' use LanguageService to submit code
     obSAS.LanguageService.Submit _
         "%include 'c:\temp\IOMTest.sas'; run;"
     MsgBox obSAS.LanguageService.FlushLog(100000)
     MsgBox obSAS.LanguageService.FlushList(100000)
End Sub
Private Sub cmdTest2_Click()
     'run the stored SAS program
     Dim obStoredProcessService As SAS.StoredProcessService
     Set obStoredProcessService = _
         obSAS.LanguageService.StoredProcessService
     obStoredProcessService.Repository = "file:c:\temp"
     obStoredProcessService.Execute "IOMtest", _
         "cond='sex eq ""M""'"
     MsgBox obSAS.LanguageService.FlushLog(100000)
     MsgBox obSAS.LanguageService.FlushList(1000000)
End Sub
Private Sub Form_Unload(Cancel As Integer)
     obWSMgr.Workspaces.RemoveWorkspaceByUUID _
     obSAS.UniqueIdentifier
     obSAS.Close
End Sub
```

When the form is loaded, the program creates a new SAS Workspace on the server using the CreateWorkspaceByServer method of the SASWorkspaceManager object. Since the connection protocol type is not specified, the default is to create a COM connection (or in this case DCOM, since the client is not running on the same host as the server).

Changing this form to use a workspace server on the Linux host, using the *IOM Bridge for COM* is as simple as modifying two of the properties of the server definition object; also, the IOM Bridge requires a valid user name and password, while the COM does not.

Example 10.2 Visual Basic Code to Open an IOM Bridge Workspace Connection

```
' create Workspace server using IOM Bridge for COM
Dim obServer As New SASWorkspaceManager.ServerDef
obServer.MachineDNSName = "hygelac"
obServer.Protocol = ProtocolBridge
obServer.Port = 8591
Set obSAS = obWSMgr.Workspaces.CreateWorkspaceByServer _
     ("", VisibilityProcess, obServer, _
      "sassrv", "sasuser", xmlInfo)
```

It is that easy to move your code from a Windows host to a Linux one. There is one important difference, however. A Windows metadata server will automatically provide logical workspace server connections as needed, but on a Linux host the object spawner has to be started manually. The process for this is documented in the SAS Technical Support note at "Quick Start: Object Spawner and Workspace Server with a Basic SAS 9 Foundation Install" at http://support.sas.com/techsup/technote/ts721.pdf.

Note that SAS is not installed on the client system—just Visual Studio. The following SAS program is stored on and run by the server:

Example 10.3 Sample SAS Program

```
%let cond=;
*ProcessBody;
proc print data=sashelp.class;
title "Test IOM Connection";
where &cond;
run;
```

The comment line `*ProcessBody;` is necessary whenever parameters are being used. In this case, the macro variable COND is initialized to null; the value of this variable can be supplied as a parameter if the program is run as a stored process.

The first button in Example10.1 (`cmdTest1_Click`) uses the SAS LanguageService interface to submit a simple SAS program. If the first subprogram is run (by clicking the **Connection** button), the **WHERE** statement is read just as `where;` which SAS ignores, so all 19 observations in the source data set are printed.

A screen capture of the first message box shown in Display 10.4 displays the SAS log generated by the server process.

Display 10.4 SAS Log from a Stored Process

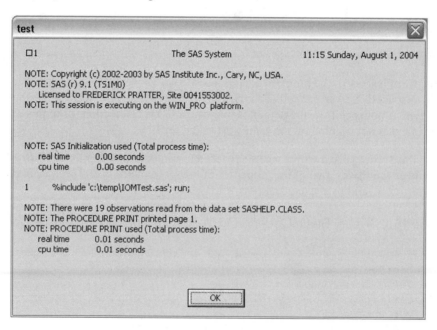

The second subprogram (`cmdTest2_click`—the **Stored Process** button event handler) uses the execute method of the StoredProcessService object, instead of the submit method of LanguageService. This code passes a name/value pair to the stored process. Here the name of the parameter is COND and the value is `sex eq "M"`; thus only the male students are listed (see Display 10.5 for the SAS log). Note that the line numbers continue from the first log, since SAS considers this a single workspace session. The form unload method then closes the workspace.

Display 10.5 SAS Log from a Stored Process with a Supplied Parameter

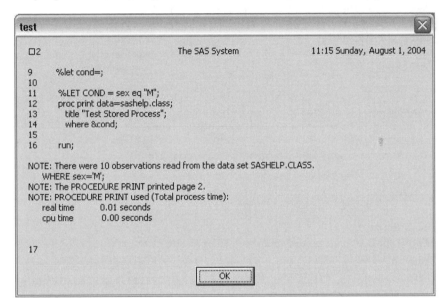

Visual C++ Interface

The COM interface and the IOM Bridge for COM work in a very similar way in Visual C++. The following code illustrates exactly the same program, ported to C++. (This code also works in Visual C++ .NET.) The program instantiates a Workspace Manager object and assigns the required properties, as follows:

Example 10.4 C++ Code to Test a DCOM Workspace Connection

```
#include <iostream>
#include <stdexcept>
#include <windows.h>

using namespace std;
#import "C:\Program Files\SAS Institute\Shared Files\Integration
Technologies\sas.tlb"
#import "C:\Program Files\SAS Institute\Shared Files\Integration
Technologies\SASWMan.dll"
int main()
{
   SASWorkspaceManager::IWorkspaceManager2Ptr pIWorkspaceManager;
SASWorkspaceManager::IServerDef2Ptr pIServerDef = NULL;
SAS::IWorkspacePtr pIWorkspace;
BSTR xmlInfo;
HRESULT hr = CoInitialize(NULL);
```

```
hr = pIWorkspaceManager.CreateInstance(
  "SASWorkspaceManager.WorkspaceManager.1");
pIServerDef.CreateInstance("SASWorkspaceManager.ServerDef");
pIServerDef->PutMachineDNSName("hygelac");
pIServerDef->Protocol = SASWorkspaceManager::ProtocolBridge;
pIServerDef->put_Port(8591);
pIWorkspace = pIWorkspaceManager->
Workspaces->CreateWorkspaceByServer(
  _bstr_t(""),                          //workspace name
  SASWorkspaceManager::VisibilityProcess,
  pIServerDef,                          // server
  _bstr_t("sassrv"),                    // login
  _bstr_t("sasuser"),                   // password
  &xmlInfo                              // connection log
);

pIWorkspace->LanguageService->Submit(
  "%include '/home/sasadm/IOMTest.sas'; run;");
MessageBox(NULL,
  pIWorkspace->LanguageService->FlushLog(10000),
  "SAS Log",
  MB_OK
);

MessageBox(NULL,
  pIWorkspace->LanguageService->FlushList(10000),
  "List Output",
  MB_OK
);

pIWorkspace->Close();
return(0);

}
```

Note the parallels to the Visual Basic code. If you have been paying attention, it is obvious that this program uses the SAS IOM Bridge for COM to connect to a Linux host. The one absolutely critical line in this program is the initialization of the COM library by the statement:

```
HRESULT hr = CoInitialize(NULL);
```

Without this line of code, your application program will abort; see the Microsoft Knowledge Base Article 169496 "INFO: Using ActiveX Data Objects (ADO) via #import in VC++" for the details of this vitally important initial step.

Since the output of this program is identical to Display 10.5, it is not shown here.

Java Client Interface

"Programmers by nature are inherently willing to trade simplicity for control: the price of control is always more effort and increased complexity" (Cooper 1999, p.96). The SAS implementation of the IOM Bridge for Java is a classic example of this approach. You can do almost anything with it, if you are willing to make the investment in learning the API. The Java program shown as in Example 10.5 is about as simple as it can be and still do something meaningful.

The application opens a connection to a remote server and runs the same SAS program as in the previous example. It will run on any client that has the SAS Java Foundation classes installed; this particular example was compiled on a Linux workstation. (This code is taken from the example "Connecting With Directly Supplied Server Attributes" in the *SAS Integration Technologies: Developer's Guide*.)

Example 10.5 Java Code to Test an IOM Bridge Workspace Connection

```
import com.sas.services.connection.Server;
import com.sas.services.connection.BridgeServer;
import com.sas.services.connection.ConnectionFactoryConfiguration;
import com.sas.services.connection.ConnectionFactoryManager;
import com.sas.services.connection.ConnectionFactoryInterface;
import com.sas.services.connection.ConnectionFactoryException;
import com.sas.services.connection.ConnectionInterface;
import com.sas.services.connection.ManualConnectionFactoryConfiguration;
import com.sas.iom.SAS.IWorkspace;
import com.sas.iom.SAS.IWorkspaceHelper;
import com.sas.iom.SAS.ILanguageService;
import com.sas.iom.SAS.ILanguageServicePackage.CarriageControlSeqHolder;
import com.sas.iom.SAS.ILanguageServicePackage.LineTypeSeqHolder;
import com.sas.iom.SASIOMDefs.GenericError;
import com.sas.iom.SASIOMDefs.StringSeqHolder;
import javax.swing.JOptionPane;

public class IOMTest{

  public IOMTest() throws ConnectionFactoryException, GenericError
  {
// connection parameters
String classID = Server.CLSID_SAS;
String host = "hunding";
int port = 8591;
String userName = "sassrv";
String password = "sasuser";

// identify the IOM Bridge server (the Workspace server)
Server server = new BridgeServer(classID,host,port);

// make a manual connection factory configuration
ConnectionFactoryConfiguration cxfConfig =
  new ManualConnectionFactoryConfiguration(server);

// get a connection factory manager
ConnectionFactoryManager cxfManager =
  new ConnectionFactoryManager();

// get a connection factory interface from the manager
ConnectionFactoryInterface cxf = cxfManager.getFactory(cxfConfig);

// get a connection from the interface
ConnectionInterface cx =  cxf.getConnection(userName,password);

// create a workspace by "narrowing" connection to the ORB
IWorkspace iWorkspace = IWorkspaceHelper.narrow(     cx.getObject()
);
```

```
// Submit batch SAS code
ILanguageService sasLanguage = iWorkspace.LanguageService();
sasLanguage.Submit("%include 'c:\\temp\\IOMtest.sas'; run;");

        // flush log file to string array
        StringSeqHolder logHldr = new StringSeqHolder();
        sasLanguage.FlushLogLines(
                Integer.MAX_VALUE,
                new CarriageControlSeqHolder(),
                new LineTypeSeqHolder(),
                logHldr);

        // display log file
        String[] logLines = logHldr.value;
        JOptionPane.showMessageDialog(null, logLines);

        // flush list file to string array
        StringSeqHolder listHldr = new StringSeqHolder();
        sasLanguage.FlushListLines(
                Integer.MAX_VALUE,
                new CarriageControlSeqHolder(),
                new LineTypeSeqHolder(),
                listHldr);

        // display list file
        String[] listLines = listHldr.value;
        JOptionPane.showMessageDialog(null, listLines);
        iWorkspace.Close();
        cx.close();
    }

    public static void main(String args[]) {

        try {
                new IOMTest();
                System.exit(0);
        }
        catch(Exception ex) {
                ex.printStackTrace();
                System.exit(1);
        }
    }
}
```

In order to run the code, it is necessary to add the required SAS IOM objects to the Java classpath. The SAS Foundation Services archive (JAR) files are installed by default in the directory **SASFoundationServices\1.1\jars** under the SAS root directory. Since SAS is not installed on the Linux client (it is not currently available for this particular flavor of Linux), the single required service connection JAR file was simply copied to the run-time directory on the client. The code to compile the sample program is as follows:

```
javac -classpath .:../sas.svc.connection.jar IOMTest.java
```

The same classpath was used to run the example. The following connection log shown is sent to the standard error output (*stderr*), presumably in order to avoid interactions with the list output. It illustrates the messages and responses sent back and forth over the TCP/IP connection to the IOM Bridge:

Display 10.6 Connection Log from IOM Bridge

```
Aug 1, 2004 12:43:52 PM com.sas.services.connection.AggregationKernel
getConnection
INFO: connection request received
Aug 1, 2004 12:43:53 PM com.sas.services.connection.BridgeServer
connect
INFO: properties for new connection: {encryptionContent=all,
port=8591, clsid=440196d4-90f0-11d0-9f41-00a024bb830c,
logFile=java.util.logging.Logger@1125127, encryptionPolicy=none,
userName=sassrv, password=xxxxxxxx, protocol=bridge, host=hunding}
Aug 1, 2004 12:43:55 PM com.sas.services.connection.AggregationKernel
getConnection
INFO: request served by unshared connection #0
Aug 1, 2004 12:44:04 PM com.sas.services.connection.AggregationKernel
reactivateConnection
INFO: connection #0 returned to factory by user
Aug 1, 2004 12:44:04 PM com.sas.services.connection.AggregationKernel
destroyConnection
INFO: connection #0 destroyed
```

The log window displayed by Java is shown in Display 10.7:

Display 10.7 Simple Java Example: Log Output

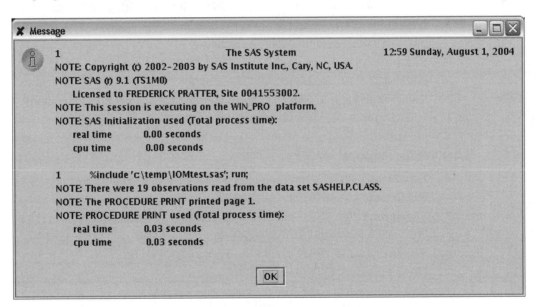

This is precisely the equivalent of the Visual Basic output shown in Display 10.4.

Conclusion

SAS Integration Technologies offers a way to connect to the flexibility and power of SAS from thin-client workstations. As Peter Eberhardt points out: "Integration technologies moved SAS from a closed, proprietary system out into the open systems area" (Eberhardt 2003). The revolutionary implications of this have not yet spread widely to the SAS user community, but an entirely new paradigm for SAS development has now become possible. At the same time, starting without SAS Integration Technologies and the Integrated Object Model is very much like those old text-based adventure games that begin: "You are in a cave. There is a key on the floor in front of you. The cave has passages to the E, W, and S. What do you want to do?" The correct answer is of course "Pick up the key."

References

SAS Technical Documentation
URL references are current as of the date of publication, but as always please check the Enterprise Integration Community pages at http://support.sas.com for the most up-to-date information.

- *SAS 9.1.2 Metadata Server: Setup Guide* – http://support.sas.com/rnd/eai/openmeta/
- *SAS 9.1 Open Metadata Architecture: Best Practices Guide* – http://support.sas.com/rnd/eai/openmeta/
- SAS Integration Technologies: Version 9 Documentation – http://support.sas.com/rnd/itech/library/library9.html
- SAS OnlineDoc (includes links to all the separate product documentation) – http://support.sas.com/91doc/docMainpage.jsp

SAS White Papers
- Jahn, Daniel. 2004. "Developing an Open Client in Visual Basic." http://support.sas.com/rnd/itech/papers/
- SAS Institute Inc. "SAS Integration Technologies: A Roadmap." http://support.sas.com/rnd/itech/papers/
- SAS Institute Inc. "SAS Integration Technologies Overview." http://support.sas.com/rnd/itech/papers/
- SAS Institute Inc. "SAS Stored Processes — A Roadmap." http://support.sas.com/rnd/itech/papers/

Publications
- Cohen, Barry R. 2004. "Using AppDev Studio and Integration Technologies for an Easy and Seamless Interface between Java and Server-Side SAS." *Proceedings of the Twenty-Ninth Annual SAS Users Group International Conference.* Cary, NC: SAS Institute Inc.
- Cooper, Alan. 1999. *The Inmates are Running the Asylum: Why High-Tech Products Drive us Crazy and How to Restore the Sanity.* 1st ed. Indianapolis, IN: Sams Publishing.

- Eberhardt, Peter. 2002. "Rev up Your Spreadsheets With Some V8 Power." *Proceedings of the Twenty-Seventh Annual SAS Users Group International Conference.* Cary, NC: SAS Institute Inc.

- Eberhardt, Peter. 2003. "SAS in the Office. IT Works." *Proceedings of the Twenty-Eighth Annual SAS Users Group International Conference.* Cary, NC: SAS Institute Inc.

- Eberhardt, Peter. 2004. "Bring the Data Warehouse to the Office with SAS Integration Technologies." *Proceedings of the Twenty-Ninth Annual SAS Users Group International Conference.* Cary, NC: SAS Institute Inc.

- Essam, Katie. 2003. *Building Windows Front Ends to SAS Software.* Leafield, Oxfordshire, UK: Amadeus Software Limited.

- Nicklin, Clare A. 2003. *The Use of Java with the SAS System.* Leafield, Oxfordshire, UK: Amadeus Software Limited.

- Nicklin, Clare A., and Daniel Morris. 2003. "A Successful Implementation of a Complicated Web-Based Application Through webAF and SAS Integration Technologies." *Proceedings of the Twenty-Eighth Annual SAS Users Group International Conference.* Cary, NC: SAS Institute Inc.

- Nipko, Joseph. "Using SAS Integration Technologies To Interface With Enterprise Applications Written In Visual C++." Scottsdale, AZ: Qualex Consulting Services.

- Pope, Lynn. 2004. *Communicating with SAS Using SAS Integration Technologies.* Leafield, Oxfordshire, UK: Amadeus Software Limited.

- Silva, Greg. 2003. "Using IOM and Visual Basic in SAS Program Development." *Proceedings of the Twenty-Eighth Annual SAS Users Group International Conference.* Cary, NC: SAS Institute Inc.

- Vodicka, Scott. 2000. "Enterprise Integration Technologies: What is it and what can it do for me?" *Proceedings of the Twenty-Fifth Annual SAS Users Group International Conference.* Cary, NC: SAS Institute Inc.

Chapter 11

Building Web Applications with SAS and Java

Overview: Java Web Applications and SAS Stored Processes

Chapter 9, "Developing Java Server-Side Applications with webAF Software," illustrated how to create servlets and JavaServer Pages using the SAS custom tags supplied with SAS AppDev Studio 3.1. This chapter describes two different approaches to Web programming using SAS Stored Processes. The SAS Stored Process Web Application, new in SAS®9, can be used by developers with little or no Java programming experience; it can be considered a replacement for the SAS/IntrNet product for some sites. The other approach using the SAS Foundation Services API offers a flexible and powerful way to build J2EE Web applications, but it is considerably more complex and requires substantial Web server and Java experience to implement.

To review, a *Java Web application* is a collection of HTML forms, JavaServer Pages, and servlets. As Chapter 8, "Java Servlets and JavaServer Pages," described, in order to provide Web content, these Java components require a servlet engine (sometimes referred to as a servlet container). Apache Tomcat is currently the most widely used platform, but compatible servlet engines are also available from BEA (WebLogic) and IBM (WebSphere). There are also a number of other open source Web container projects, such as Jetty and Winstone, but these have not been validated to work with SAS.

A *SAS Stored Process* is simply a SAS program that has been registered in the SAS Metadata Server so that it can be accessed from another application, such as SAS Enterprise Guide or the SAS Stored Process Web Application; see the preceding chapter for more information about defining SAS servers in the SAS Open Metadata Architecture.

At present, there are three ways to use SAS Stored Processes:

1. *IOM Direct Interface Stored Processes* were introduced in SAS 8 and operate only on a SAS Workspace Server. This approach is deprecated in SAS®9, although it continues to be supported.

2. *The SAS Stored Process Web Application* is an alternative to the SAS/IntrNet Application Broker; like the earlier approach it requires no Java programming experience on the part of the developer.

3. The *Stored Process Service* application programming interface (API) can be used to run stored processes, either from JavaServer Pages, servlets, custom tagsets and/or Java applications. The Stored Process Service API requires SAS Foundation Services.

Information on the first of these is available from the *SAS Integration Technologies: Developer's Guide*. The rest of this chapter is about the two new components introduced in SAS®9, the SAS Stored Process Web Application and the Stored Process Service API.

Note that a stored process service does not always require a stored process server. The two things are quite different, although it is not always obvious how. A stored process server is a component of the SAS Open Metadata Architecture in SAS®9. Technically, it is a logical server defined in the metadata to provide load balancing and shared multi-user access to SAS data and procedures. In contrast, a stored process service is implemented through a set of Java classes that allow an application program to access SAS Stored Processes. A stored process service can use either a workspace server or a stored process server.

Using the SAS Stored Process Web Application as a Replacement for SAS/IntrNet

As noted in Chapter 7, "SAS IntrNet: htmSQL," many Web developers have been moving away from the older CGI technology that was the basis for SAS/IntrNet, toward Microsoft .NET or cross-platform Java Web applications. The SAS Stored Process Web Application potentially can replace SAS/IntrNet at some sites. Using this approach requires installing the SAS Web Infrastructure Kit, currently available on Windows, Solaris (64-bit), AIX (64-bit) and HP-UX IPF. Unfortunately, it is not yet available for Linux, but as always, you need to check with SAS to determine what components are available for your site.

A further issue is that Microsoft Internet Information Server (IIS) is not compatible with Java applications and does not support servlet containers. You have to install Tomcat or another servlet engine alongside IIS in order to use the SAS Stored Process Web Application. If you want to install the Apache HTTP server, it must be on a different TCP port from IIS, since there can be only one listener on port 80 at a time. Finally, the new technology requires a substantial effort by the SAS system administrator to set up; sites without the luxury of a full-time administrator may not want to convert yet.

The advantages of the new software, however, are significant enough that sites where it is possible to make the transition should do so, at least for new applications. The first part of this chapter, then, explains the procedures that are necessary to install the new software. Once this has been accomplished, creating Web programs that use SAS Stored Processes is relatively easy, and does provide performance and security enhancements that may be worth the effort. For more information about converting from SAS/IntrNet to the SAS Stored Process Web application, see the section "Converting SAS/IntrNet Programs to Stored Processes" in the *SAS Integration Technologies: Developer's Guide*.

The SAS Stored Process Web Application

In order to create a SAS Stored Process Web Application it is necessary first to install two new SAS Integration Technologies components: the SAS Web Infrastructure Kit and SAS Foundation Services. (This is documented in the SAS *Integration Technologies: Developer's Guide* under "Stored Process Software Requirements," although curiously it is not mentioned in the SAS Integration Technologies *SAS Web Infrastructure Kit* documentation.) In SAS 9.1 these two products are available from the CDs labeled "SAS Client-Side Components Volume 2," although in this context they are not client-side components at all but middle-tier elements that should be installed on the server host.

Installing the SAS Web Infrastructure Kit

You also should install *SAS Management Console 9.1* software on a client workstation with access to the SAS server. In addition, *SAS Enterprise Guide 3.1* software is handy for creating and testing stored processes, although not essential. Two other required third-party software components are the correct version of the Java Development Kit (JDK) and a validated servlet container such as Tomcat 4.1.

The SAS 9.1.3 installation media should include two CDs labeled "Third Party Software Components":

Volume 1 contains the *Java Software Development Kit Version 1.4.2_04* validated for use with the SAS components. (Note that as of this writing, this release of the JDK is no longer available from Sun. The SAS installation instructions further suggest that you turn off automatic updates for the product.)

Volume 2 has links to validated releases of two other products: Tomcat Version 4.1.18 and Apache Web Server 2.0.45. These three components (the JDK, the servlet engine, and optionally the Web server) should be installed before the SAS products. Volume 2 also has links to the BEA and IBM servlet container download pages, if these are preferred to Apache Tomcat.

The following table, based on the "Stored Process Software Requirements" section in the *SAS Integration Technologies: Developer's Guide*, summarizes the required software configuration:

Table 11.1 Stored Process Software Requirements

Software	Server	Client
Servlet container (e.g., Tomcat 4.1)	Required	
HTTP server (Apache 2)	Optional	
SAS Web Infrastructure Kit (WIK)	Required	
SAS Foundation Services	Required	Optional
Base SAS, SAS/GRAPH, SAS Integration Technologies Version 9.1	Required	
Java Development Kit (JDK 1.4.2)	Required	Optional
SAS Management Console 9.1	Required on either server or client	Required on either server or client
SAS Enterprise Guide		Optional

First, install Tomcat (or one of the other two approved servlet engines) on the host. The Apache Web server is optional, since Tomcat can also serve HTML as well as JSP and servlets.

Next install Base SAS, the correct JDK (not Java 1.5), SAS Foundation Services, and the SAS Web Infrastructure Kit, in that order, on the server host. SAS Management Console software can be installed on the server or on a network client. If the server is in a remote location, it is much more convenient to have SAS Management Console on your desktop system. As noted above, you probably want to have SAS Enterprise Guide 3.1 software on the client desktop, since it will make the process of developing and testing stored processes much simpler.

Finally, if you are going to be developing Java programs to use the Stored Process Service API, you will need the correct JDK and SAS Foundation Services on your desktop. If you have SAS AppDev Studio 3.1 installed on the client, webAF software can be used to create Java Web applications using the SAS Foundation Services classes (see the second part of this chapter for examples).

The SAS Integration Technologies installation procedure has been improved considerably in SAS 9.1.3, but it is still a laborious and time-consuming job. Table 11.2 lists the nine SAS Integration Technologies components on the "SAS Client-Side Components Volume 2" installation CD. This table is provided to clarify the somewhat intimidating selection of installation choices. Presumably, this arrangement will change in future releases, but this table is nonetheless useful if only as a guide to the platforms for which these components are currently available and where to find documentation on them; see http://support.sas.com/rnd/itech/library/library9.html for the most current list.

Table 11.2 SAS Integration Technologies Components: Installation

SAS Integration Technologies Documentation
Platforms: Windows, 64-bit Enabled Solaris, 64-bit Enabled AIX, HP-UX IPF, z/OS, 64-bit Enabled HP-UX, Linux, Tru64 UNIX
Description: Installs a local copy of the online documentation. Includes the following volumes, plus a glossary of terms:
Documentation:
<ul><li>*Technical Overview*</li><li>*Server Administrator's Guide*</li><li>*Administrator's Guide*</li><li>*Administrator's Guide (LDAP Version)*</li><li>*Developer's Guide*</li><li>*SAS Web Infrastructure Kit: Overview*</li><li>*SAS Web Infrastructure Kit: Administrator's Guide*</li><li>*SAS Web Infrastructure Kit: Developer's Guide*</li></ul>
SAS Web Infrastructure Kit 1.0
Platforms: Windows, Solaris (64-bit), AIX (64-bit), HP-UX IPF
Description: Creates Web applications to access SAS data and services; requires a J2EE servlet container such as Apache Tomcat, BEA WebLogic or IBM WebSphere.
Documentation:
<ul><li>*Technical Overview: SAS Stored Processes*</li><li>*Administrator's Guide: SAS Stored Processes*</li><li>*Developer's Guide: SAS Stored Processes*</li><li>*SAS Web Infrastructure Kit: Administrator's Guide*</li><li>*SAS Web Infrastructure Kit: Developer's Guide*</li></ul>

SAS Foundation Services 1.1

Platforms: Windows, 64-bit Enabled Solaris, 64-bit Enabled AIX, HP-UX IPF, 64-bit Enabled HP-UX, Linux

Description: Includes a set of Java classes to provide the following services:
- metadata and content repository access
- stored process execution
- IOM client connection service for Java
- dynamic service discovery
- user authentication
- profile management
- session context management
- activity logging
- event management
- information publishing

Documentation:

- *Technical Overview: Foundation Services*
- *Administrator's Guide: SAS Foundation Services*
- *Developer's Guide: SAS Foundation Services*
- *Developer's Guide: Java Clients*
- *Developer's Guide: Foundation Services API*

SAS BI Web Services for .NET 1.3

Platforms: Windows only

Description: Provides Web services for the .NET framework; used to invoke and list SAS Stored Processes. SAS BI Web Services require a connection to a SAS Metadata Server and a SAS Stored Process Server.

Documentation:

- *Technical Overview: SAS BI Web Services*
- *Developer's Guide: SAS BI Web Services*

SAS BI Web Services for Java 1.0

Platforms: Windows, 64-bit Enabled Solaris, 64-bit Enabled AIX, HP-UX IPF

Description: Provides cross-platform Web services for Java; used to invoke and list SAS Stored Processes. SAS BI Web Services require a connection to a SAS Metadata Server and a SAS Stored Process Server.

Documentation:

- *Technical Overview: SAS BI Web Services*
- *Developer's Guide: SAS BI Web Services*

SAS Integration Technologies Client for Windows 9.1
Platforms: Windows only **Description:** Windows application to communicate with a SAS server; installed automatically as part of the Base SAS or SAS Integration Technologies client installation, this executable is required only if the SAS Integration Technologies client is installed by itself.
SAS Integration Technologies Administrator 1.6
Platforms: Windows only **Description:** Java application to manage object servers, spawners and the publishing framework; requires an installed LDAP server. **Documentation:** *Administrator's Guide (LDAP Version): Getting Started* – "Using the Integration Technologies Administrator"
SAS Package Reader 1.6
Platforms: Windows, 64-bit Enabled Solaris, 64-bit Enabled AIX, HP-UX IPF, 64-bit Enabled HP-UX, Linux, Tru64 UNIX **Description:** Client application to retrieve contents of a SAS package as an archive file, usually as an attachment to an e-mail message. An archive is denoted by an .spk file extension, which is an abbreviation for "SAS Package." **Documentation:** *Developer's Guide: Publishing Framework* – "SAS Package Reader"
SAS Subscription Manager 1.5
Platforms: Windows, 64-bit Enabled Solaris, 64-bit Enabled AIX **Description:** Java applet that interacts with an LDAP server to provide services to manage subscriptions; the documentation indicates that this applet will not be supported in future releases of SAS Integration Technologies. **Documentation:** *Developer's Guide: Publishing Framework* – "SAS Subscription Manager"

Configuring the Server

The only tricky part of all this is installing the SAS Web Infrastructure Kit (WIK). The pre-installation checklist for setting up required user accounts indicates that six specific local accounts should be created on the metadata server. You can customize these for your site, but these are the defaults:

- SAS Administrator (sasadm)
- SAS Demo User (sasdemo)
- SAS General Server (sassrv)
- SAS Guest (sasguest)
- SAS Trusted User (sastrust)
- SAS Web Administrator (saswbadm)

Make sure to create these account on the Windows host system first, before installing the WIK—you will need to specify them as you are installing the software. After the WIK has been installed successfully, there should be a file called **instructions.html** in the installation directory **\SAS\9.1**, specifying the manual configuration steps required to complete the deployment.

A sample set of steps is illustrated in Display 11.1; the actual steps you will see are based on the specific products licensed for your site and the options provided during the initial installation.

Throughout this document, this image indicates that an automation script has been provided to execute the steps of this section. After executing any of the automated scripts, you may need to refresh the view in SAS Management Console in order to see the updates.

High-Level Overview of the Steps

Note: The steps listed below were generated based on the software that you planned to configure on this machine.

Display 11.1 Web Infrastructure Kit Configuration

1. Start your Metadata Server
2. Start the SAS Management Console
3. Define your Metadata Repository
4. Define your Metadata Users
5. Defining Default Authorizations
6. Define your SAS Application Server
7. Define your Stored Process Server
8. Define your OLAP Server
9. Define your Data Step Batch Server
10. Define your Object Spawner
11. Define your SAS/CONNECT Server
12. Define your SAS/CONNECT Spawner
13. Define your SAS/SHARE Server
14. Define your Job Scheduler Server
15. Define your HTTP Server
16. Define the SAS Foundation Services to the metadata
17. Load SAS Stored Process samples
18. Load Web Infrastructure Kit "primer" metadata
19. Start your Object Spawner
20. Start your OLAP Server
21. Start your SAS/CONNECT Spawner
22. Start your SAS/SHARE Server
23. Start your SAS Services Application
24. Deploying your Web Applications

25. Start your Tomcat Server

26. Using your Applications

27. Getting More Information

The biggest change between SAS 9.1 and SAS 9.1.3 is the introduction of the batch scripts, indicated by the ⟳ symbol. Assuming that you have performed a generic installation and that all of the software has been set up correctly, the automation scripts should simplify the process substantially.

In addition to the configuration instructions, a second information file is supplied under the SAS home directory in **Web\Portal2.0.1\wik_readme.html**. This document provides additional instructions for installing the SAS Web Infrastructure Kit and contains information for setting up security for the different servlet engines.

The most important issue in the WIK **readme** file relates to environment variables and how Tomcat is installed. On a Windows host running Windows 2000, XP or some other NT descendant, Tomcat can be run as a background service that is loaded automatically when the system boots up. The Tomcat 4.1 installation procedure for Windows includes several optional check boxes. By default, **Install as NT Service** is not checked. If you select this check box (and it is recommended) after the installation you will have to remove the Tomcat service and replace it; the setup instructions are included in the **wik_readme.html** instructions.

First, to uninstall the service, type the following at a command prompt window:

```
%CATALINA_HOME%/bin/tomcat.exe -uninstall "Apache Tomcat 4.1"
```

Next, reinstall with the options specified by SAS; these can be implemented as a batch file as shown in Example 11.1 below.

Example 11.1 Tomcat Installation Batch File

```
set JAVA_HOME=C:\j2sdk1.4.2_05
set CATALINA_HOME=C:\Tomcat4.1
set CATALINA_OPTS=-Xms512m -Xmx1024m -server
XX:-UseOnStackReplacement -Djava.awt.headless=true

rem The following command should be on a single line
%CATALINA_HOME%/bin/tomcat.exe
install Apache-Catalina %JAVA_HOME%/jre/bin/server/jvm.dll -
Djava.security.manager
Djava.security.policy=%CATALINA_HOME%/conf/catalina.policy -
Djava.class.path=%CATALINA_HOME%/bin/bootstrap.jar;
%JAVA_HOME%/lib/tools.jar
Dcatalina.home=%CATALINA_HOME% %CATALINA_OPTS% -Xrs
start org.apache.catalina.startup.BootstrapService
params start
stop org.apache.catalina.startup.BootstrapService
params stop
out %CATALINA_HOME%/logs/stdout.log
err %CATALINA_HOME%/logs/stderr.log\
```

If you do not install Tomcat as a service, then you will need to restart the Tomcat servlet engine every time the host reboots. Also, you cannot change your mind and run the preceding batch file later, since the Tomcat 4.1 executable is only loaded if you install it initially as a service.

To start the servlet engine manually, use **Start ▶ Programs ▶ SAS ▶ 9.1 ▶ Start Tomcat** rather than **Start ▶ Programs ▶ Apache Tomcat 4.1 ▶ Start Tomcat,** since the SAS installed batch startup file specifies some additional functions that are required for the SAS Stored Process Web Application. In order to stop Tomcat, use the default selection **Start ▶ Programs ▶ Apache Tomcat 4.1 ▶ Stop Tomcat.** Assuming everything is now installed correctly, it is time to write a SAS Stored Process.

Using SAS Stored Processes

In order to illustrate how to convert a SAS/IntrNet program to use the SAS Stored Process Web Application, we shall return to an example from Chapter 6, "SAS/IntrNet: the Application Dispatcher," which illustrated how to generate dynamic output using the SAS/IntrNet Application Broker and ODS. The relevant portion of the sample program shown in the earlier chapter is reproduced as Example 11.2.

For this example, the program is stored as **shoes.sas** in the **My Documents** folder for user sas on the remote host.

Example 11.2 Sample Program to Generate Dynamic Output

```
%* Sales report Example - Display Product by Region;
%macro salesrpt;

%global region;
proc report data=sashelp.shoes;
by region;
%if (&region ne ) %then %do;
where region="&region";
%end;
title "Sales by Product by Region";
footnote "Data are current as of &systime &sysdate9";
column product sales;
define product / group;
define sales / analysis sum;
quit;
%mend salesrpt;
%salesrpt
```

The steps required to build and run a stored process are documented in the *SAS Integration Technologies: Developer's Guide.* The following example assumes that you have installed SAS Enterprise Guide 3.1 software to create and test the SAS program and you have installed SAS Management Console to register the SAS program as a stored process. Note that there is also a different Stored Process Wizard in SAS Enterprise Guide software; you can use one or the other to register the stored process, depending on which way is more convenient.

Creating a Stored Process Repository with SAS Management Console 9.1

The SAS Web Infrastructure Kit installation procedure installs a default stored process repository called *Foundation* along with a set of sample processes. Unless you want to use **/Stored Process/Sample** as the URL for all of your Web applications, you will need to create a new folder using the SAS Management Console Stored Process Manager.

Display 11.2 SAS Management Console Stored Process Manager

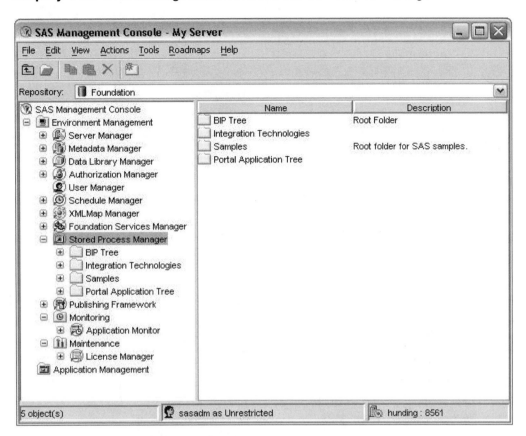

Display 11.2 shows the initial configuration for the Stored Process Manager. To create a new
folder, select **Actions ▶ New Folder** or just click the new folder icon on the toolbar. A single
screen appears, on which a name for the new repository folder can be supplied—for this example,
the new folder is `BBU Examples`. Note that this is a *logical* folder; it need not exist as a physical
directory on the host.

As noted above, this example uses the sample code from Chapter 6. In order to convert this to a
stored process using SAS Management Console, it is necessary add three lines of code, as shown
in Display 11.3 in bold. (If you use the SAS Enterprise Guide Stored Process Wizard, the
additional macros are added to the program automatically.)

The comment `ProcessBody` at the top indicates the beginning of the stored process. This
comment is required for this program, since the program uses macro variables as parameters. The
two macro invocations `%stpbegin` and `%stpend` call corresponding built-in macros that supply
parameters to the Web application and ODS. Most of these macro variable values can be
overridden to produce different output effects as desired; see "Stored Processes: Input
Parameters" in the *SAS Integration Technologies: Developer's Guide* for the details.

Display 11.3 SAS Stored Process Program

```
*ProcessBody;
%global  region;
%stpbegin;
%* Sales report Example - Display Product by Region;
%macro salesrpt;

proc report data=sashelp.shoes;
by region;
%if (&region ne ) %then %do;
where region="&region";
%end;
title "Sales by Product";
footnote "Data are current as of &systime &sysdate9";
column product sales;
define product / group;
define sales / analysis sum;
quit;
%mend salesrpt;
%salesrpt
%stpend;
```

To start the SAS Management Console Stored Process Wizard, right-click the new folder and select **New Stored Process**.

Display 11.4 Create New Stored Process

Supply a name for the process, as shown. This name will be used in the URL for the Web application. The description and keywords are optional. Click **Next** and the screen shown in Display 11.5 should appear.

Display 11.5 Specify Server, Source File and Result Type in the New Stored
Process Wizard

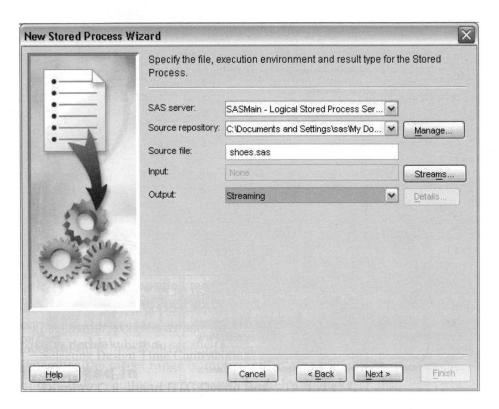

In order to register the stored process, you must supply four values:

1. **SAS server** – Select either the stored process server or the workspace server. In this case, we
    want to use the former so that the Web application is shareable and the results are sent back to
    the client. The workspace server cannot be used to stream output in HTML format, although it
    can send the output to a file.

2. **Source repository** – Select the physical location where the SAS program file is installed. It can
    be in any location that the stored process server can access and does not have to be on the same
    physical machine as the SAS server. To add a new location, select **Manage**.

3. **Source file** – Specify the SAS program, in this case the shoes.sas data set.

4. **Output** – For a Web application, select **Streaming**. The default is **None**, which will result in
    the program running with no output, which is almost certainly what you do not want to happen.

The **Input** field is used to get XML or other streaming input data and is not needed for this
example.

Display 11.6 Add Run-time Parameters in the New Stored Process Wizard

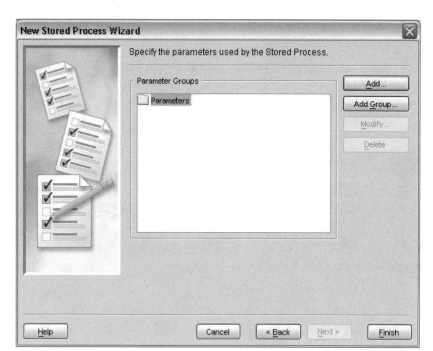

The screen shown in Display 11.6 is used to specify the macro parameters that will be supplied at run time. Click **Add** to add this parameter and the screen shown in Display 11.7 should appear.

Display 11.7 The Add Parameter Dialog Box

Typing the word **region** into the **Label** field and pressing the tab key automatically causes the same value to appear in the **SAS variable name** field. Click **OK**, and then click **Finish** to create and register the new stored process.

Testing the Stored Process

In order to check whether the new stored process works as desired, you can open SAS Enterprise Guide 3.1 software and select **View ▶ Stored Process List**.

Display 11.8 SAS Enterprise Guide Stored Process List

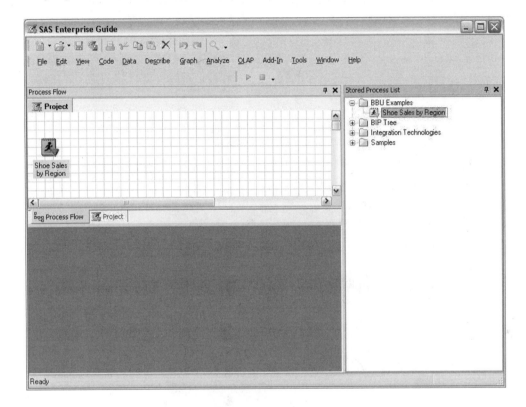

Drag **Shoe Sales by Region** to the Process Flow window or right-click the icon in the Stored Process List and select **Add to Project**.

Right-click the icon in the Process Flow window and select **Run This Stored Process**. Since the sample program requires a region parameter, a window will pop up asking for the value of this run-time parameter. Typing **Asia** as shown in Display 11.9 runs the report for this region. Note that the footnote uses the built-in macro variables &sysdate and &systime, but there is no need to specify these values.

Display 11.9 Prompt for Run-time Parameter

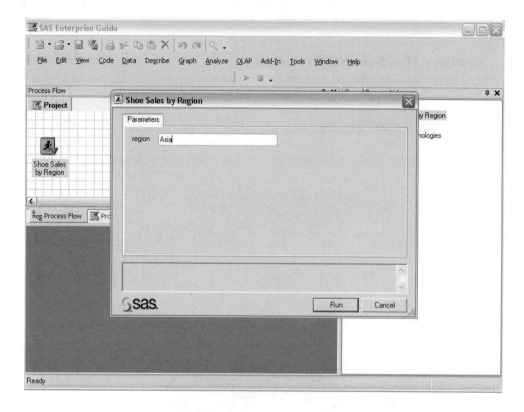

At this point a connection is made to the stored process server, which requires a valid SAS user ID to run, as shown in Display 11.10.

Display 11.10 Enter SAS User ID and Password

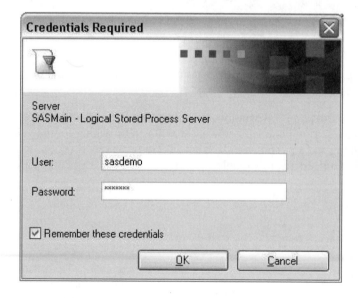

Note that this can be any of the user accounts defined in the installation. Once a valid user ID and password have been supplied, the stored program should run and display the results for the specified region, as Display 11.11 illustrates.

Display 11.11 SAS Enterprise Guide: HTML Streaming Output

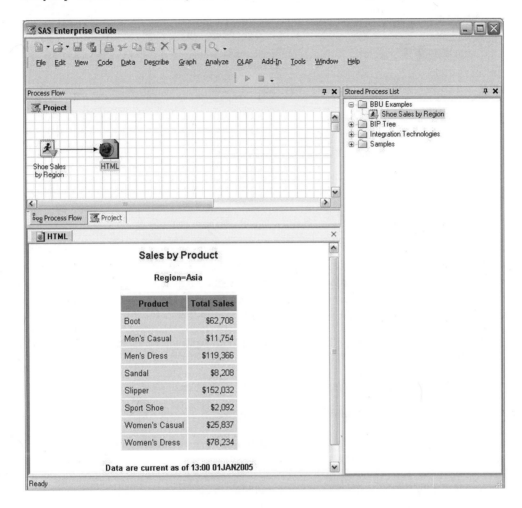

Using the SAS Stored Process Web Application

Now that the program has been successfully debugged, running it as a Web application is simply a matter of specifying the correct URL in your favorite Web browser window. Since the repository folder is **BBU Examples** and the name of the stored process is **Shoe Sales by Region**, the corresponding URL as shown in Display 11.12 is:

```
http://hunding:8080/SASStoredProcess/do?
    program=/BBU%20Examples/Shoe%20Sales%20by%20Region&region=Asia
```

Note that spaces in the URL are replaced by the equivalent ASCII %20, which codes for a space character.

Display 11.12 HTML Output from Web Application

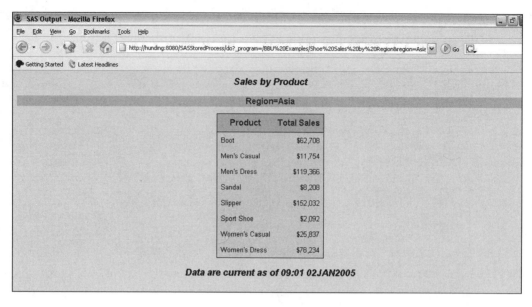

If the Apache Web server was installed prior to the WIK configuration, the HTTPD server is set up to automatically forward requests to the servlet container, so it is not necessary to specify the Web server port. In this case, since Tomcat is running as the only Web server, the required server address is http://hunding:8080; port 8080 is the default port for this Web container.

The command to run the SAS Stored Process Web Application is very similar to that used with SAS/IntrNet for the same function:

SAS/Intrnet Application Broker: `/cgi-bin/broker`

Stored Process Web Application: `/SASStoredProcess/do`

As is usual for HTML, the run-time parameters follow the question mark character (?) and are separated by ampersands (&). The first parameter, `_program`, is required and specifies the folder and the procedure name `/BBU Examples/Shoe Sales by Region`. Since the desired report is to be run solely for the Asia region, the second parameter is `region=Asia`.[1]

As this example illustrates, it is not necessary to write any Java code at all in order to use the SAS Stored Process Web Application. The only requirements are some SAS programming skills. As one developer noted at a recent user group conference, the SAS Stored Process Web Application is "the Son of IntrNet." The section that follows, on the other hand, is intended for developers who are comfortable with Java and JSPs and who would like to write Web applications using the Stored Process Service API.

---

[1] Note that quotes around the value of an HTML parameter must be URL-encoded; see http://httpd.apache.org/info/css-_security/encoding_examples.html for a general discussion of how to specify URLs using entities.

Using the Java Foundation Services Stored Process Service

A JavaServer Page or servlet that accesses the Stored Process Service API can be developed using any Java interactive development environment (IDE) such as Eclipse (see "About Us" at http://www.eclipse.org/org/) or even a text editor such as Notepad or TextPad from Helios Software Solutions (http://www.textpad.com/). The following example uses webAF software, since it provides powerful features for debugging and testing Java code.[2]

Managing Configuration Issues

The Stored Process Service can be used to build Web applications that are deployed to a servlet container. Consequently, the software requirements for the Stored Process Service are similar to those for the SAS Stored Process Web Application. The SAS Web Infrastructure Kit 1.0 must be installed on the Web server in a qualified servlet container such as Tomcat 4.1. In order to compile the Java code on the client, SAS Foundation Services 1.1 must be installed there as well.

One way to deploy a Web application using SAS Foundation Services is to use the SAS Remote Services Application, which is installed as part of the SAS Web Infrastructure Kit. The file `C:\Program Files\SAS\Web\Portal2.0.1\configure_wik.bat` configures the Remote Services Application on the server; it should add a link on the **Start ▶ Programs ▶ SAS ▶ 9.1** menu to the startup file `\SAS\9.1\Lev1\web\Deployments\RemoteServices\ WEB-INF\StartRemoteServices.bat`. The SAS Remote Services Application starts a listener on port 5099 by default. At present, there is no way to install this as a Windows service, so it should be restarted after each server reboot.

Finally, the file `C:\Tomcat4.1\conf\catalina.policy` contains the permissions for Java programs to run within the servlet container. As noted in the following section, it is necessary to modify this file to add explicit permissions for each Web application deployed to the server.

Creating a SAS Stored Process Web Application with Foundation Services

Currently the easiest way to create a new SAS Stored Process Web Application is with webAF software, although as previously noted it can be created in any Java IDE. In the latter case, make sure to modify the classpath to include the required Java archive files, installed by default in `\sas\SASFoundationServices\1.1\jars`. SAS AppDev Studio 3.1 software also provides a built in servlet container, running on port 8082 on the local host, which can be used for testing the resulting application, as shown in the following procedure.

The steps for creating the sample application in webAF software are as follows:

1. Open webAF software and create a new Web application. For this example, call the application `SASFoundationServices`.

2. At Step 2 of the New Project Wizard, choose **Blank Web Application**.

3. At Step 3, unselect the check box for **SAS Taglib and Tbeans (Version 3)**, as these will not be needed for this example.

4. At Step 4, select the initial content as **Servlet**. Name the servlet `SimpleServlet`.

---

[2] Note that SAS AppDev Studio 3.2 will not include support for webAF software but instead uses the Eclipse open-source product as the development environment. Nonetheless, the discussion of how to create a Stored Process Service should be correct in outline.

5. At Steps 5 through 8, accept the defaults by clicking **Next**. This will create the new HTTP servlet.

6. In the webAF code window, replace the automatically generated **SimpleServlet.java** program with the code shown in Example 11.3.

Example 11.3 Sample Stored Process Service Application[3]

```
package servlets;
import java.io.*;
import javax.servlet.*;
import javax.servlet.http.*;

import com.sas.services.discovery.LocalDiscoveryServiceInterface;
import com.sas.services.discovery.DiscoveryService;
import com.sas.services.discovery.ServiceTemplate;
import com.sas.services.user.UserServiceInterface;
import com.sas.services.user.UserContextInterface;
import com.sas.services.session.SessionServiceInterface;
import com.sas.services.session.SessionContextInterface;
import com.sas.services.storedprocess.StoredProcessServiceFactory;
import com.sas.services.storedprocess.StoredProcessServiceInterface;
import com.sas.services.storedprocess.StoredProcessInterface;
import com.sas.services.storedprocess.ExecutionInterface;
import com.sas.services.connection.BridgeServer;
import com.sas.services.connection.Server;
import com.sas.services.connection.ConnectionFactoryInterface;
import com.sas.services.connection.ConnectionFactoryManager;
import com.sas.services.connection.ConnectionInterface;
import com.sas.services.connection.ConnectionFactoryConfiguration;
import com.sas.services.connection.ManualConnectionFactoryConfiguration;
import com.sas.services.deployment.MetadataSourceInterface;
import com.sas.services.deployment.OMRMetadataSource;
import com.sas.services.deployment.ServiceLoader;

public class SimpleServlet extends HttpServlet
{
public void doPost (HttpServletRequest request,
  HttpServletResponse response) throws IOException
{
PrintWriter out=response.getWriter();
response.setContentType("text/html");

// run-time values for Metadata server connection parameters
String host="hunding";
String port = "8561";
String userName = "sasdemo";
String password = "password";
String repository = "Foundation";
String softwareComponent = "Remote Services";
String serviceComponent = "BIP Remote Services OMR";

// run-time values for Stored Process server connection
int bridgeport = 8611;
String file="c:\\Documents and Settings\\sas\\My Documents";
```

---

[3] Thanks to David Barron at SAS for figuring out this example.

```
String pgm="shoes.sas";
   try {
       // connect to Metadata server on port 8561 to discover Services
       LocalDiscoveryServiceInterface discoveryService =
               DiscoveryService.defaultInstance();
       MetadataSourceInterface metadataSource =
               new OMRMetadataSource(
               host,port,userName,password,repository,
               softwareComponent,serviceComponent);
       ServiceLoader.lookupRemoteDiscoveryServices(
               metadataSource, discoveryService);

       // create user context
       ServiceTemplate stp =      new ServiceTemplate(
       new Class[] {UserServiceInterface.class} );
       UserServiceInterface userService =
               (UserServiceInterface)
               discoveryService.findService(stp);
       UserContextInterface user =
               userService.newUser(userName,password,"DefaultAuth");

       // create session context
       stp = new ServiceTemplate( new Class[]
               {SessionServiceInterface.class} );
       SessionServiceInterface sessionService =
               (SessionServiceInterface)
               discoveryService.findService(stp);
       SessionContextInterface sessionContext =
               sessionService.newSessionContext(user);

       // create stored process service
       StoredProcessServiceFactory spFactory =
               new StoredProcessServiceFactory();
       StoredProcessServiceInterface spServiceInterface =
               spFactory.getStoredProcessService();
       StoredProcessInterface spi =
               spServiceInterface.newStoredProcess(
               sessionContext,
               StoredProcessInterface.SERVER_TYPE_STOREDPROCESS,
               StoredProcessInterface.RESULT_TYPE_STREAM);

       // send messages to stored process
       spi.setSourceFromFile(file,pgm);
       spi.setParameterValue("region","Canada");
       spi.addInputStream("_WEBOUT");

       // connect to Stored Process Server on
       //     load balancing port 8611
       BridgeServer server = new BridgeServer(
               Server.CLSID_SASSTP,host,bridgeport);
       ConnectionFactoryConfiguration cxfConfig =
               new ManualConnectionFactoryConfiguration(server);
       ConnectionFactoryInterface cxf =
               ConnectionFactoryManager.
               getConnectionFactory(cxfConfig);
       ConnectionInterface ci =
       cxf.getConnection(userName,password);
```

```
        // run stored process
        ExecutionInterface ei = spi.execute(false,null,false,ci);

        // display results
        InputStream is = ei.getInputStream("_WEBOUT");
        BufferedReader br = new BufferedReader(
                new InputStreamReader(is));
        String temp = "";
        while((temp = br.readLine()) != null) {
                out.println(temp);
        }
    }
    catch (Exception ex) {
        out.println(
                "<html><body>" + "SAS encountered an error: " +
                ex.getLocalizedMessage() + "</body></html>");
    }
}

public void doGet(HttpServletRequest request,
    HttpServletResponse response)
    throws ServletException, IOException
{

    doPost(request, response);
}
}
```

This example is by no means a complete introduction to using the Java Foundation Services. The complete API is documented in the *SAS Integration Technologies: Developer's Guide* under "SAS Foundation Services" in the section on "Java Clients." The documentation also includes a link to the Foundation Services API for the Java packages currently available from SAS. The following brief overview is provided simply as a guide to the workings of this particular example.

For this example, the run-time values are supplied as constants. In a production application, they would more properly be stored as a resource bundle and accessed as session properties. Note that the values for the stored process assume that the program **shoes.sas** (described in the preceding section on the SAS Stored Process Web Application) has been registered as a stored process with SAS Management Console.

This program uses two OMA servers. First, a connection is made to the metadata server to find the stored process registry. Then a second connection is made to the stored process server in order to create the Web output.

The first connection takes place in three steps:

1. Get the default discovery service.
2. Specify the metadata source.
3. Use the Service Loader to look up the remote discovery service on the host.

Once the remote services are discovered, the user context and the session context must be created in order for the stored process service to run. A stored process service can now be instantiated using the defined session context and the stored process server. Methods are available in the interface to specify the location of the SAS program on the server, pass parameters such as region=Canada, and to add the default _webout destination.

In practice, the region parameter would not be hard-coded into the servlet but would be passed from the HTTP request object. Note that with this code, it is not possible to specify a parameter to the stored process on the URL; it must be specified using the object's setParameterValue method.

The second connection is then made to the load-balancing server, configured using SAS Management Console on the default port 8611. The program uses the Connection Factory class to connect to the server, as illustrated in Chapter 10, "Using the SAS Open Metadata Architecture with the Integrated Object Model."

To run the stored process, the execute method of the object is called with four arguments:

- `boolean synchronous`
- `ExecutionStatusListenerInterface listener`
- `boolean createAlert`
- `Object connection`

Since this stored process generates streaming output, the generated HTML can be captured by reading from the file `_webout` by an ordinary BufferedReader. It can then be redirected to the servlet reponse object.

Note that in order for this application to run as a Tomcat servlet, the Foundation Services JAR files should be copied to the **webapps/WEB-INF/lib** directory under the SAS AppDev Studio project directory. It is a "feature" of Tomcat 4.1 that all of the required classes must be available in each of the application home directories; it is not sufficient for them to be available from the Java `classpath` environment variable. The SAS AppDev Studio software installation process should have copied these files to the correct location, but they can be copied over manually from the **\sas\SASFoundationServices\1.1\jars** folder if necessary.

To test the sample application in webAF, the following steps are required:

1. Select **Build (F7)** from the Build toolbar.
2. From the **Tools** menu, select **Start Java Web Server**.
3. Click **Execute in browser** from the Build toolbar.
4. If all goes well, the Web page shown in Display 11.13 should be displayed as http://localhost:8082/SASFoundationServices/SimpleServlet.

Display 11.13 Sample Stored Process Service Output

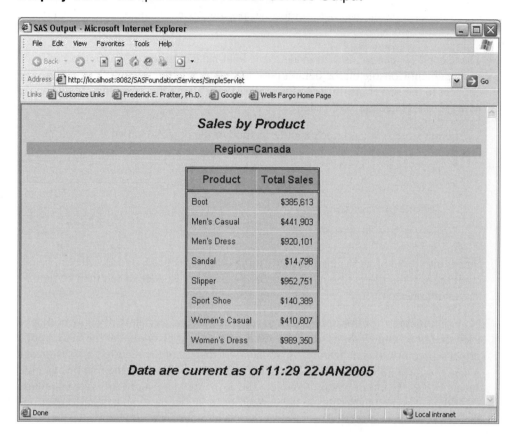

Deploying a SAS Stored Process Web Application

As previously noted, this application is running on the local servlet container provided with SAS AppDev Studio 3.1 software on the client. In order to move the application to a remote host, a Web Application Archive (WAR) file is necessary. The procedure for moving an application to another server is described in Chapter 9. To review, the webAF Package Wizard provides an easy way to create a deployment file from a project.[4]

When you select the wizard from the **Tools ▶ Wizards** menu, you should see something like the following:

---

[4] You can also use the WAR tool in Eclipse or the Ant build tool to make WAR files.

Display 11.14 WAR File Package Wizard

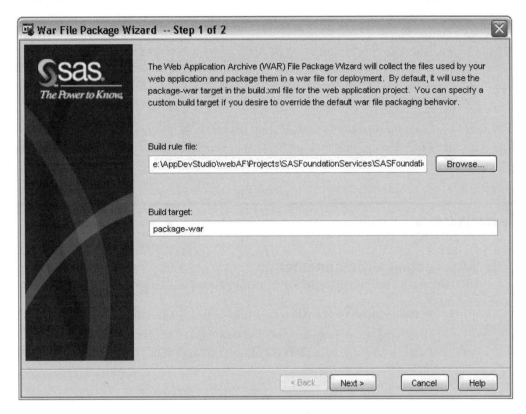

Click **Next** and then click **Finish** on Step 2. The WAR file will be created in the project directory, in this case as **\AppDevStudio\webAF\Projects\SASFoundationServices\ SASFoundationServices.war**. Copy the WAR file to the Tomcat **webapps** directory on the server.

Before attempting to run the application, edit the server **catalina.policy** file to add the appropriate permissions. These can be fine-tuned at will (if you know how to write Java security files) but the following code, which allows all permissions, can be used for building and testing the application.

```
grant codeBase "file:${catalina.home}/webapps/SASFoundationServices/-"
{
   permission java.security.AllPermission;
};
```

Now restart Tomcat from the **Services** menu or use the batch startup file to load the new application. In this case, the name of the remote host is hunding, so typing the URL http://hunding:8080/SASFoundationServices/SimpleServlet should display the same result as shown in Display 11.13.

Conclusion

SAS Stored Processes can be used to automate repetitive tasks and to supply dynamic Web content. The SAS Stored Process Web Application was designed for SAS developers who are not familiar with Java and who may need to update SAS/IntrNet applications. The main advantage of using stored processes is the integration with the SAS Open Metadata Architecture, which provides better management and security. The Stored Process Service API, on the other hand, was designed for Java developers, who may not be familiar with SAS, to embed SAS remote data and computing services in Java Web applications. In either case, SAS has supplied a comprehensive set of tools that can be used to generate dynamic server-side Web pages.

References

SAS Technical Documentation
URL references are current as of the date of publication.

- Foundation Services - API – http://support.sas.com/rnd/gendoc/bi/api/Foundation/

- SAS Integration Technologies: Version 9 Documentation at http://support.sas.com/rnd/itech/library/library9.html includes the following online manuals:

  □ *SAS Integration Technologies: Server Administrator's Guide*

  □ *SAS Integration Technologies: Administrator's Guide*

  □ *SAS Integration Technologies: Administrator's Guide (LDAP Version)*

  □ *SAS Integration Technologies: Developer's Guide*

  □ *SAS Web Infrastructure Kit 1.0: Overview*

  □ *SAS Web Infrastructure Kit 1.0: Administrator's Guide*

  □ *SAS Web Infrastructure Kit 1.0: Developer's Guide*

SAS White Papers
- SAS Institute Inc. "SAS Stored Processes — A Roadmap." http://support.sas.com/rnd/itech/papers/

Part 5

Appendixes

Appendix A

DHTML: Dynamic Programming on the Web Client with JavaScript

The primary focus of this book has been on the various ways to create Web pages in which dynamic content is generated on the *server* using the Common Gateway Interface (Part 2) and JavaServer Pages (Parts 3 and 4). This appendix is a general overview of an alternative strategy for developing Web programs on the *client* using Dynamic HTML (DHTML). Appendix B introduces Java applets and presents a number of examples illustrating how to create and deploy Java applets using webAF software.

In general, *Dynamic HTML* refers to Web content that can change each time it is viewed. The term is usually employed to describe methods for producing interactive content on the client using a combination of JavaScript and style sheets. Note that this is only one of two definitions of DHTML. Many Web developers use the term DHTML to refer to the *Document Object Model* (DOM) standard from the World Wide Web Consortium (W3C). As the W3C Web site points out:

> "Dynamic HTML" is a term used by some vendors to describe the combination of HTML, style sheets and scripts that enables documents to be animated. The W3C has received several submissions from members companies on the way in which the object model of HTML documents should be exposed to scripts. These submissions do not propose any new HTML tags or style sheet technology. The W3C DOM WG is working hard to make sure interoperable and scripting-language neutral solutions are agreed upon.
> (http://www.w3.org/DOM)

Consequently, in this book the first definition of DHTML is assumed—that is, a "combination of HTML, style sheets and scripts."

Although this is a widely used approach, there are some significant disadvantages to using JavaScript for Web development. Foremost among these is the problem of browser incompatibilities. A large proportion of the effort required to deploy DHTML pages is just to make sure the programs run on most browsers. The most common browsers—Internet Explorer, Netscape Navigator, Mozilla Firefox, and Opera—all support (or do not support) different JavaScript features. In addition, scripting problems can be caused by different versions of the same browser (Internet Explorer 5.0 vs. Internet Explorer 5.5, Netscape 6 vs. Netscape 4), as well as by different platforms (PC, Macintosh, Palm). A great deal of ingenuity has been devoted to dealing with this problem, but ultimately unless the Web designer knows specifically what kind of clients to support, DHTML is not necessarily an effective solution to the problem of providing dynamic content.

The following introduction is by no means intended to be comprehensive. There are many good tutorials available on creating DHTML with JavaScript; several of these are included in the references at the end of this chapter. The intent of this discussion is simply to provide a framework for comparing this technique with the other approaches available.

JavaScript Is Not Java

The original version of the Netscape Web browser included some basic scripting capabilities called LiveScript. When Navigator 2.0 came out, it included the ability to run Java applets, so Netscape decided to rename the scripting language JavaScript, in order to capitalize on the popularity of the new Java programming language. In fact, the two are quite different creatures: Java is a complete object-oriented programming language, while JavaScript is an object-based scripting language that can be used only from within a Web browser. The main difference is that JavaScript does not support all of the features of Java. From an end-user standpoint that just means JavaScript is a lot easier for non-programmers to learn; see http://mozilla.org/js/ for detailed information about the JavaScript implementation.

Microsoft implemented its own scripting language for the Internet Explorer browser called *JScript*, which is roughly compatible with JavaScript.[1] In June 1997, at the joint request of Netscape and Microsoft, an international standards body called the European Computer Manufacturers Association (ECMA) produced the ECMA-262 standard, charmingly referred to as *ECMAScript*. This standard was intended to resolve the differences between the two different approaches from Netscape and Microsoft. Internet Explorer 6.0, Mozilla Firefox, and later browsers all more-or-less support the new standard. The browser incompatibility problems noted previously should go away as older, non-ECMA versions disappear, but at present the Web developer must always be conscious that there are numerous versions of JavaScript available and not all scripts will work in every browser.

---

[1] Note that Microsoft Active Server Pages can make use of JScript for providing server-side content; this should not be confused with using JavaScript on the client.

The big advantage that client-side scripts executed by the Web browser have over server-side programs is that the former can provide dynamic content without waiting for a message to and from the server. As a consequence, JavaScript is widely used for applications where response speed is critical. For example, *rollovers* (when an image changes as you move the cursor over it) are almost always coded in JavaScript. A rollover requires replacing an image on a Web page after the image has already loaded. You can even download all of your graphics to the client before any rollovers occur. In this way images appear very quickly, since the browser already has the required content. This is generally an effectve use of JavaScript, even though not all browsers support rollovers.

Another good use for JavaScript is data validation. For example, an HTML form can be used to check for the presence of required fields. If something is missing, the browser script can display a message and prompt the user for the input needed, without waiting for a round trip to the server. A related use is password verification. Many login pages require that a password be entered twice; a JavaScript can easily compare the two values and determine if they are equal. If not, the user can be prompted to re-enter the text. Of course, you cannot do password authentication in the browser. It would be a pretty big security risk if a site allowed users to validate their own passwords!

As noted above, JavaScript is based on something called the Document Object Model (DOM). This model is a hierarchical way to represent the objects created by the Web browser. The W3C established a standard called DOM Level 1 in 1998. This standard is not entirely compatible with the older object models used by Netscape and Microsoft. The most recent standard, DOM Level 3, was published in 2004; see http://www.w3.org/DOM/. W3C DOM will probably be used in future browsers, but for the moment Web developers must deal with the fact that there are three different versions of the Document Object Model for JavaScript: Microsoft, Netscape and the W3C.

The important thing to realize about DOM is that it is intended to be language neutral. It just happens to work in JavaScript. As the W3C Document Object Model Activity Statement puts it:

> W3C's Document Object Model (DOM) is a standard Application Programming Interface (API) to the structure of documents. The DOM aims to make it easy for programmers to access components and to delete, add, or edit their content, attributes and style. In essence, the DOM makes it possible for programmers to write applications which work properly on all browsers and servers and on all platforms.... (http://www.w3.org/DOM/Activity)

Note that work on the Document Object Model is currently suspended in favor of a new project called *Rich Web Clients*; see http://www.w3.org/2006/rwc/Activity.html for a description of the newer proposal. Nevertheless, the discussion that follows is not likely to be superseded in the near future.

A DOM provides standardized tags for all of the objects on a Web page. The map describes a hierarchy of objects, starting with *window* as the top level; the objects on the page are all *children* of the window object, and in turn they can have child objects of their own. For example, the following script could be used to display the current date:

Example A.1 JavaScript Date Example

```
<html>
<head>
  <title>Display Date Using JavaScript</title>
</head>
<body>
  <script type="text/javascript">
            today = new Date();
        var message = "<h1>Today is " +
                  (today.getMonth()+1) + "/" +
        today.getDate() + "/" +
        today.getYear() + ".</h1>";
            document.write(message);
  </script>
</body>
</html>
```

The output of this page in Firefox 1.0 and Internet Explorer 6.0 is shown in the following displays:

Display A.1 HTML Output: Firefox 1.0

Display A.2 HTML Output: Internet Explorer 6.0

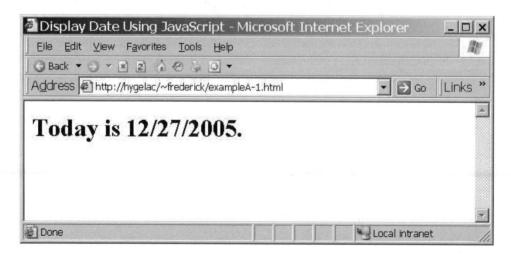

Note that the `getYear()` JavaScript function in Firefox returns a three-digit rather than a four-digit year. The same script produces two different results, depending on the choice of Web browser!

The JavaScript is enclosed in `<script></script>` tags to separate it from page content; without these the text of the script would show up, rather than the results. The `<script>` tag has one required attribute, `type`. (Older browsers support the `language` attribute, but this has been deprecated in favor of `type`.) Several different kinds of scripting languages are available; this example demonstrates how to indicate that you are using JavaScript. Note that attributes are always surrounded by double quotes.

The `Date` object is one of the many JavaScript built-in classes. A new instance of this object, called `today` is created by the following statement:

```
today = new Date();
```

The object is initialized by default using the computer's system clock. A variable named `message` is declared and assigned the date values, using the `getMonth()`, `getDate()` and `getYear()` methods of the `Date` object. Finally, the `write()` method of the `document` object is called to display the resulting string to the browser window.

JavaScript code looks a little like a cross between SAS and Visual Basic. Statements are separated by semicolons. (Although that's not strictly necessary; it's considered good form.) JavaScript variables have no types, so you don't need to worry about converting numeric data to character. The "dot" notation for specifying the methods and properties of objects is similar to the way these are implemented in Java, C++, and Visual Basic. Finally, the whole thing is embedded in an HTML page, so that the Web browser can display it.

The following example is slightly more complex. This page uses JavaScript to convert between the Fahrenheit and Centigrade temperature scales. The script is called from an HTML `form` object that collects and displays the information.

Example A.2 JavaScript Temperature Conversion Calculator

```html
<html>
<head>
  <title>JavaScript Temperature Conversion Calculator</title>

  <script type="text/javascript">
  function calc(n) {
      var temp = document.calculator.input.value;
      if (n == 1)
              document.calculator.result.value = 5*(temp-32)/9;
              else
              document.calculator.result.value = (9*temp)/5+32;
  }
  </script>
</head>
<body>
<div style="text-align: center">
<h1 style="color: blue">Temperature Conversion Calculator</h1>
<form name="calculator">
  <p><strong>Enter a temperature and
        select a conversion type: </strong>
```

```
      <input type="text" name="input" /></p>
      <p><strong>Result: </strong>
      <input type="text" name="result" /></p>
      <p><input type="button" onclick="calc(1)"
          value="Fahrenheit to Centigrade" /></p>
      <p><input type="button" onclick="calc(2)"
          value="Centigrade to Fahrenheit" /></p>
</form>
</div>
</body>
</html>
```

The output for this program in Firefox is shown in Display A.3.

Display A.3 JavaScript Temperature Conversion Form

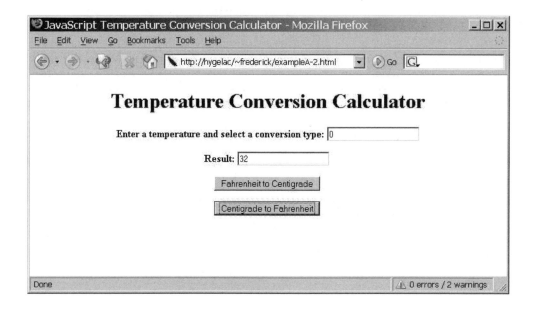

The form contains four input controls: two text boxes and two buttons. The temperature value 0° C is entered in the first text box. Clicking the **Centigrade to Fahrenheit** button runs the JavaScript program and displays the result as 32° F. In the script, the input temperature text box is referenced as `document.calculator.input`—that is, the text box named `input` on the form named `calculator` in the current document.

This example also makes use of JavaScript's *event handler* mechanism. Event handlers are programs that respond to user actions. In this case, when the user clicks one of the buttons it triggers an `onclick` event, which calls the script with the parameter 1 for Fahrenheit to Centigrade, and 2 for Centigrade to Fahrenheit, depending on which button was clicked.

In summary, JavaScript is a useful mechanism for developing dynamic Web clients. With careful attention to browser incompatibilities, Java Scripts can be very efficient for many kinds of applications. The reader is cautioned, however, against an over reliance on this approach, since the practice is widely overused and is likely to be detrimental to good user interface design. The best strategy is to use server-side pages where these can be implemented efficiently, and to reserve scripting for applications that are best done on the client. For example, users are frequently prompted to enter their e-mail address or password twice, for validation. A typical DHTML function might compare the two values and display an alert if they do not match. Checking the password against a list of valid users must of course be done on the server, since downloading the entire password list to the client would represent an unacceptable security risk.

A p p e n d i x **B**

Client-Side Programming with Java Applets

Using Java Applets

The most common alternative to JavaScript for client-side dynamic HTML is to use Java *applets*. An applet is a Java program that runs in the Web browser window. The most obvious difference between JavaScript and Java is that in scripting languages like JavaScript, the Web designer is using HTML to control the display. In applets, on the other hand, there is no `<form>` tag. The user interface is generated by Java, using the methods built in to the `Applet` class.

Java is a large and complex language that is growing every year. It is not terribly hard to learn, but it does take a lot of practice; it would not be possible to include even an introductory tutorial here. Sun no longer provides their applet tutorial on their Web site, presumably because they feel the useful life of this technology is ending; however, they do still provide some sample applet code at http://java.sun.com/developer/codesamples/ applets.html. For more help on applets, see one of the basic tutorials available on the Web, such as http://www.realapplets.com/tutorial/.

The SAS HTML Formatting Tools macros (described in Chapter 3, "Creating Static HTML Output,") use Java applets to display SAS/GRAPH output in the browser window. It is not necessary to be a Java programmer to run these applets, but the following discussion may be useful for users who want more information about this process. The best that can be hoped for here is to present an overview of how applets work, without focusing too much on why they work that way.

Example B.1 illustrates the HTML code to include and run a Java applet in Microsoft Internet Explorer 6:

Example B.1 Java Temperature Conversion Applet

```
<html>
<head>
  <title>Java Temperature Conversion Applet</title>
</head>
<body>
  <div style="text-align: center">
  <h1 style="color: blue">
    Temperature Conversion Calculator</h1>
  <object code="Calculator.class"
      width = 480
      height = 120
      alt="Java Applets are not supported. You need to update
        your browser!" />
  </div>
</body>
</html>
```

The important thing to note about this example is that all of the formatting and calculations are provided by the applet. The HTML document includes only the page title and a heading; everything else is supplied by the Java program.

In general, running a program in Java is a two-step process. First, the Java program source code must be compiled, using the `javac` compiler available free from Sun Microsystems as part of the Java Software Development Kit (JDK). There are various versions of the JDK available; the current one is known both as *Java 2* and *JDK 1.5*. (Sun likes to use two different numbering systems for versions in order to keep programmers from getting bored and restless; Java versions 1.0 and 1.1 were just known as "Java," but since version 1.2 the product is now called "Java 2.")

It is important to note that SAS AppDev Studio applications were written for JDK 1.4.2. It is possible to run two versions of the JDK on the development system. As noted in Chapter 9, "Developing Java Server-Side Applications with webAF Software," if you have a more recent version of Java on your development system, it is probably best to install JDK 1.4.2_04, which is still available for download from Sun at http://java.sun.com/products/archive/. The resulting applets should work in most versions of the Java Runtime Environment (JRE), but again, the problem with client-side scripting is that there is no way to know whether they will run correctly on different client configurations.

The applet code is called by the `code="calculator.class"` attribute on the `<object>` tag. Early versions of Java used the `<applet>` tag to execute Web programs on the client. In HTML 4.0, the `<object>` tag was introduced as an alternative; they both do the same thing, but the W3C felt that naming the tag "object" better identified its use.

The source code for the applet must be saved as a file with the extension .java; in the previous example the program is called **calculator.java**. Compiling this program produces a file called **calculator.class**. Class files in Java are saved as something called "byte code"; that is, the Java program statements (which can be read and hopefully written by humans) are transformed into a program that can be read by the computer.

Most browsers have the capability to execute Java class files. The actual applet program is the class file, not the original Java source code. Java byte code is executed by the *Java Virtual Machine* (JVM). This is not really a machine, but a program that runs on different computers. When you download the Java JDK, the JVM is automatically installed, or you can install the JVM separately as the *Java Runtime Environment* (JRE).

Each kind of system—Windows, UNIX, Linux, Macintosh or Palm—has a JVM that is specific to that system. In this way, you can create byte code on one computer—for example a Windows PC—and run it on another, completely different platform, such as a handheld device. This makes Java applets very attractive. The developer does not need to know what kind of system the user has, since the same byte code can be run on any system that supports Java. This "write once, run anywhere" concept is one of the reasons Java has become so popular for Web development.[1]

In the preceding example, the class file is stored in the same directory as the HTML file that displays it. In fact, the class file can be stored in another directory of the same computer, or even on another computer altogether. If this is not possible in your environment, the `codebase` and `archive` attributes of the `<applet>` tag can be used to specify where to find the class file. For this example, the class file and the HTML file are both stored in the same public folder on the Web server, so these tags are not required.

At this point, you are probably wondering what the Java source code for this example looks like. Here it is:

Example B.2 Temperature Conversion Applet

```
// Java Temperature Conversion Applet
import java.applet.*;
import java.awt.*;
import java.awt.event.ActionListener;
import java.awt.event.ActionEvent;

public class calculator
   extends Applet
   implements ActionListener
{
   private TextField tfInput, tfResult;
   private Button btF2C, btC2F;
```

---

[1] In actuality, the matter is not so simple. Some people have argued that the motto ought to be "write once, debug everywhere."

```
private final String F2C = new String("Fahrenheit to
   Centigrade");
private final String C2F = new String("Centigrade to
   Fahrenheit");

//initialize user interface
public void init()
{
        //set layout and foreground colors
        setLayout(new BorderLayout());
        setBackground(Color.white);
        setFont(new Font("Times New Roman", Font.PLAIN,
               14));

        //create text fields
        tfInput = new TextField(6);
        tfResult = new TextField(6);

        //set result field read only
        tfResult.setEditable(false);

        //create buttons
        btF2C = new Button(F2C);
        btC2F = new Button(C2F);

        //register listeners
        btF2C.addActionListener(this);
        btC2F.addActionListener(this);

        //create 3 panels
        Panel p1 = new Panel();
        p1.add(new Label(
"Enter a temperature and select conversion type:"));
        p1.add(tfInput);

        Panel p2 = new Panel();
        p2.add(new Label("Result:"));
        p2.add(tfResult);

        Panel p3 = new Panel();
        p3.add(btF2C);
        p3.add(btC2F);

        // add panels to frame
        add(BorderLayout.NORTH, p1);
        add(BorderLayout.CENTER, p2);
        add(BorderLayout.SOUTH, p3);
}
```

```
//event handler for buttons
public void actionPerformed(ActionEvent e)
{
        String actionCommand = e.getActionCommand();
        double t1, t2;
        // get input temp
        t1 = (Double.valueOf(tfInput.getText()))
            .doubleValue();

        // compute result
        t2 = actionCommand.equals(F2C) ? 5*(t1-32)/9 :
            9*t1/5 + 32 ;

        // display result
        tfResult.setText(String.valueOf(t2));
    }
}
```

Don't panic if it's not obvious what is going on in this program; it's not that hard to figure out. Actually, even if you are not a Java programmer, it is possible to figure out what is going on if you have a little background information. Java Version 1.1 applets used the *Abstract Windows Toolkit* (AWT). This is a package of classes and methods that can be used to create graphical programs that will display on different systems without modification. It works because each different system has some way to create form controls, such as buttons. A Windows button is produced by the Windows API, while a Linux button is displayed by an entirely different program. All of these details are handled for you by the JVM. All your need to do is include the correct AWT statement. For example, the statement

```
Button btF2C = new Button("Fahrenheit to Centigrade");
```

creates a new button object on the form, with the label specified in quotes.

It is very important to recognize the difference between coding a Web page as an applet and using JavaScript. In JavaScript, the buttons on the HTML form are created by the `<input>` tags on the form. In Java applets, on the other hand, the buttons are created by the Java AWT. You do not need to code for the buttons in HTML; you just need to include the applet in an `<object>` tag on the Web page. The disadvantage of this is that the applet must be written by a Java programmer; HTML coders cannot create pages unless they can also code in Java. The solution for this, presented later in this chapter, is to use JavaBeans, reusable software components that can be combined to develop graphical user interfaces.

As the following display demonstrates, the resulting HTML page as displayed in Internet Explorer looks very much like the JavaScript output from Appendix A, "DHTML: Dynamic Programming in the Web Client with JavaScript."

Display B.1 Temperature Conversion Form (Java Applet)

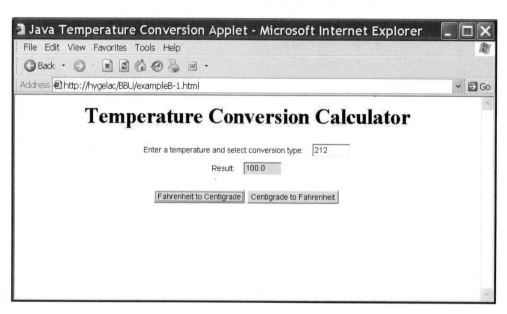

The advantage of coding the application in this way as opposed to using JavaScript is that it should work in most browsers without modification. Unfortunately, as we shall see, it's not that simple.

Getting Into the Swing

There are two big problems with the AWT, which Sun addressed in the JDK 1.2 (the first version of Java 2). Both of them relate to the way in which form controls like buttons and text boxes are generated. First, the AWT used the native code engines to display forms; a page displayed on a Windows PC will look different from the same page on Linux. This may or may not be a problem, but it does mean you have to do a lot of testing on different platforms to make sure that your applets are doing what you want them to.

The second problem is somewhat more technical. The AWT controls are so-called *heavyweight* components; rendering these is delegated to underlying native platform window controls. This means that each heavyweight component requires an operating system process. In contrast, *lightweight* graphical objects are rendered without delegating to any native platform window controls. As a result, lightweight components use fewer system resources to render an equivalent GUI.

Sun figured out how to get applets to slim down. The result was the set of *Swing* components, which are 100% pure Java. Thus they display the same way on all platforms, and they avoid the overhead associated with creating the native AWT components. (See http://java.sun.com/docs/books/tutorial/uiswing/learn/ for the Java Swing tutorial.)

The following example shows how to convert the applet in Example B.1 to Swing; only the Java source code is different. The HTML code is almost identical, except that the statement `code="Calculator.class"` has to be changed to `code="JCalculator.class"` instead. As Example B.3 illustrates, the convention is that Swing classes have the same names as the corresponding AWT class, but with a prepended "J." So this is a reasonable name for the new class.

Example B.3 Temperature Conversion Applet (Swing)

```java
import javax.swing.*;
import java.awt.*;  // include font, colors, layouts
import java.awt.event.ActionListener;
import java.awt.event.ActionEvent;

public class JCalculator
  extends JApplet
  implements ActionListener
{
  private JTextField tfInput, tfResult;
  private JButton btF2C, btC2F;
  private final String F2C = new String
      ("Fahrenheit to Centigrade");
  private final String C2F = new String
      ("Centigrade to Fahrenheit");

  //initialize user interface
  public void init()
  {
      //format the applet's content pane
      getContentPane().setLayout(new BorderLayout());
      getContentPane().setBackground(Color.white);
      getContentPane().setFont
      (new Font("Times New Roman", Font.PLAIN, 14));

  //create text fields
      tfInput = new JTextField(6);
  tfResult = new JTextField(12);

  //set result field read only
  tfResult.setEditable(false);

  //create buttons
  btF2C = new JButton(F2C);
  btC2F = new JButton(C2F);

  //register listeners
  btF2C.addActionListener(this);
  btC2F.addActionListener(this);
```

```
//create 3 panels
JPanel p1 = new JPanel();
p1.add(new JLabel
("Enter a temperature and select conversion type:"));
p1.add(tfInput);

JPanel p2 = new JPanel();
p2.add(new JLabel("Result:"));
p2.add(tfResult);

JPanel p3 = new JPanel();
p3.add(btF2C);
p3.add(btC2F);

// add panels to content pane
getContentPane().add(BorderLayout.NORTH, p1);
getContentPane().add(BorderLayout.CENTER, p2);
getContentPane().add(BorderLayout.SOUTH, p3);
}
}
```

Since the event handler method is the same as in Example B.2, it is not shown here.

Three kinds of changes were necessary to convert an applet from AWT to Swing. First, the statement

```
import java.applet.*;
```

has been replaced by

```
import javax.swing.*;
```

instead. That's because this program uses the `JApplet` class instead of `Applet`.

The second change is that all of the form components—`TextField`, `Button`, `Panel` and `Label`—have been replaced with their lightweight Swing equivalents— `JTextField`, `JButton`, `JPanel` and `JLabel`.

Finally, Swing has introduced the `ContentPane` object; all of the form components are added to this container instead of directly to the applet. (All these concepts are discussed in the Java Swing tutorial; if you are interested in more detail about these changes, you are encouraged to look there or in one of the references listed at the end of this chapter.)

The output of this program, as shown in Display B.2, looks a little different from the previous example, for an interesting reason. The original AWT used the *native* Graphical User Interface on each platform; an applet would look different depending on whether you ran it on Windows, Linux, or Macintosh. The Swing interface is 100% Java. If you want, you can customize the look and feel of your Swing applet for each type of system, but you don't have to. The default is that the page will look the same no matter which browser or platform you use to view it.

Display B.2 HTML Output: Temperature Conversion Applet (Swing)

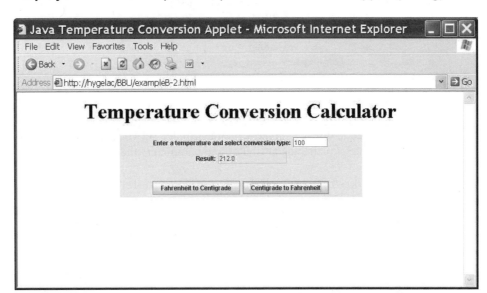

In general, the message from Sun is that you should be using the Swing components instead of the AWT. There is only one last little problem: the final critical issue regarding applets, the Java browser *plug-in.*

Getting Plugged In

In order to run Swing applets in a Web browser, you have to have one additional Java component installed. As Sun points out:

> The Java Plug-in software is a component of the Java Runtime Environment (JRE). The JRE allows applets written in the Java programming language to run inside various browsers. The Java Plug-in software is not a standalone program and cannot be installed separately.
> (http://java.com/en/download/faq/java_plugin.xml)

If you have never opened an applet in your Web browser, you will need to download and install a copy of this for each browser that you use.

It is possible to modify your HTML page as follows to automatically download the plug-in.

Example B.4 HTML Source for Locating and Downloading the Java Plug-in

```
<html>
<head>
    <title>Java Temperature Conversion Applet</title>
</head>
<body>
        <div style="text-align: center">
        <h1 style="color: blue">Temperature Conversion
            Calculator</h1>
        <object classid="clsid:CAFEEFAC-0015-0000-0000-
            ABCDEFFEDCBA"`
width="480" height="120"
```

```
        codebase="http://java.sun.com/products/plugin/autodl/
        jinstall-1_5_0-windows-i586.cab#Version=1,5,0,0">
            <param NAME="code" VALUE="JCalculator.class">
            <param NAME="codebase" VALUE="/BBU">
            <param name="scriptable" value="false">
        </object>
        </div>
    </body>
    </html>
```

This HTML code directly replaces the far simpler HTML page in Example B.1. The following table explains the attributes and parameters that appear in Example B.4:

Table B.1 Java Plug-in Attributes and Parameters

Attribute or Parameter	Description
`classid="clsid:CAFEEFAC-0014-0002-0000-ABCDEFFEDCBA"`	The name of the plug-in on Sun's Web site; attribute of `<object>`
`codebase="http://java.sun.com/products/plugin/autodl/jinstall-1_5_0-windows-i586.cab#Version=1,5,0,0">`	The location of the Java plug-in for Internet Explorer; attribute of `<object>`
`<PARAM NAME="code" VALUE="JCalculator.class">`	The name of the applet class
`<PARAM NAME="codebase" VALUE="/BBU">`	The path to the Java class file relative to the Web server document root
`<PARAM name="scriptable" value="false">`	Indicates that no JavaScript is used in the page

Opening Example B.4 in Internet Explorer results in a message asking if you want to download the plug-in. Clicking the **Yes** button will download the plug-in; on a 56KB dialup connection it can take up to half an hour to copy. Fortunately, it is only necessary to do this once, but most users will not want to wait! Clearly, this approach is only practical if all of your users are on a broadband or LAN connection, as they would be in most large organizations. Applets, clever though they may be, are not yet ready for the home consumer without a high-speed Internet connection.

One other major issue with Java applets is that the `<object>` tag only works in Internet Explorer. If the Web client browser is Mozilla or Firefox, you need instead to supply an `<embed>` tag. If you want your pages to work in a mixed environment, you have to supply both an `<object>` and an `<embed>` tag, with the latter embedded within a `<comment>` tag. Refer to the Java documentation at http://java.sun.com/j2se/1.5.0/docs/guide/plugin/developer_guide/using_tags.html for the details of using the `<applet>`, `<object>` and `<embed>` tags for cross-platform development.

Creating Java Applets with webAF Software

With SAS AppDev Studio software, you can create Java programs that access the features of SAS from within a *thin client* application. This does not necessarily mean that you have to put your users on a diet; a thin client is an application running on a system that does not have SAS installed. Most commonly this will be via a browser to render Web pages that in turn access SAS data sets and procedures. One way to do this, as we saw in Appendix A, is to create Java applets that can be dynamically loaded on the client as needed.

A Simple Applet Example

SAS supplied some simple applet examples in the SAS AppDev Studio Version 2 online documentation; see http://support.sas.com/rnd/appdev/V2/webAF/examples.htm for an easy explanation of how to create an applet in webAF software. The documentation for SAS AppDev Studio 3.1 provides a number of more complex examples (see http://support.sas.com/rnd/appdev/examples/), but as an introduction to using SAS AppDev Studio, the following discussion is a good place to start.

The following instructions have been modified slightly to bring them up to date; see Chapter 9, "Developing Java Server-Side Application with webAF Software," for more information about building webAF projects.

1. Create a new project named `HelloWorld`. Specify the project type to be `Empty Project`.
2. Add a new Java source file named **`HelloWorld.java`** to the project, and enter the Java source code (see Example B.5).
3. Save the Java source file.
   Note: Java requires that you use the same name for the source file as you use for the name of the class.
4. Compile the Java source file.
5. Add a new HTML page to the project and name it **`HelloWorld.html`**. Next, enter the HTML code (see Example B.6).
6. Save the HTML file.
7. Change the project properties. On the **Options** tab, change the **Launch applets with the following HTML file** field to point to your new **`HelloWorld.html`** file.

Note that the instructions suggest that you begin with an empty project. The SAS AppDev Studio 3.1 New Project Wizard can be used to create an applet project, but for this example the additional functionality is not necessary; see the following discussion for how to create an applet project to display an HTML form. Example B.5 shows the demo HelloWorld applet Java code.

Example B.5 HelloWorld Applet

```
import java.awt.Graphics;

public class HelloWorld extends java.applet.Applet
{
   public void paint( Graphics g )
   {
             g.drawString( "Hello, World!", 60, 30 );
   }
}
```

The associated HTML page is illustrated in Example B.6.

Example B.6 HelloWorld HTML

```
<html>
<head>
   <title>Hello World Example</title>
</head>
<body>
<hr>
<applet code="HelloWorld.class" name="HelloWorld"
   width=200 height=200 />
<hr>
</body>
</html>
```

The SAS AppDev Studio 2.0 product documentation indicates that you can use the Java *appletviewer* to display this applet; this is not supported in SAS AppDev Studio 3.1 software. You can either move the HTML and class files to a Web server directory or, in Windows XP, you can just open the page in Windows Explorer by clicking on the file icon. In this case, the name of the Web page is `C:\AppDevStudio\webAF\Projects\HelloWorld\HelloWorld.html`. Display B.3 shows the result. (Note that since the HTML page uses the `<applet>` tag and not the `<object>` tag, the applet works in Firefox.)

Display B.3 HelloWorld Applet

You can also use webAF software to edit pre-existing Web pages. For example, you can create a webAF project to display the Fahrenheit-to-Centigrade temperature conversion form. Follow these steps:

1. Start the New Project Wizard by selecting **File ▶ New ▶ Projects ▶ Empty Project**. For this example, name the project `Calculator`.

2. Select **Insert ▶ File into project** from the menu bar.

3. Add the Java source code **JCalculator.java** and the HTML file to the project. You should now be able to open the HTML page to see the resulting Web page.

There is no particular advantage to using webAF software with existing applets, but it does enable the experienced Web programmer to browse and debug Java code in a flexible and powerful integrated development environment.

Creating a New Applet in webAF Software

The Calculator project uses an HTML form with a Java applet to collect and display information. You can use webAF software to build this same project from scratch, as shown in the following procedure:

1. Select **File ▶ New Applet Project** from the webAF main menu to start the New Project Wizard. Name the new project `webAFCalculator`. Click **OK**.

2. The next screen enables you to specify the name of the HTML file to launch the applet. Leave the text box blank; the wizard will generate a page with the same name as the project.

3. On the same screen you can select whether to use Swing or AWT. As previously noted, Swing applets are less resource intensive and consequently are preferred to the older AWT approach.

4. Selecting **Swing Look-and-Feel** as **System default** results in the blank form shown in Display B.4; the other choices currently are **Java default**, **Metal**, **CDE/Motif** and **Windows**. You are encouraged to experiment with different styles to find one you like.

5. Click **Next** and then **Finish**. After some moments while the project is being created, the following screen is displayed (only the **Visuals** window is shown):

Display B.4 Initial Web Calculator Form

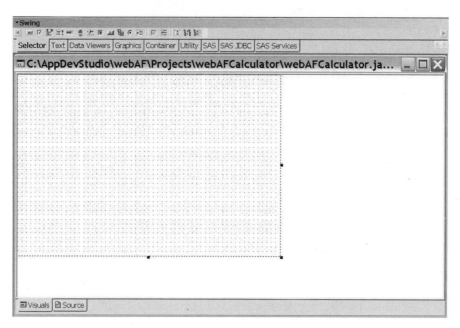

Selecting the second tab in the Project Navigator window on the left shows that three files have been generated:

- **webAFCalculator.fml** – the applet frame file
- **webAFCalculator.html** – a Web page to display the applet
- **webAFCalculator.java** – the applet source code

6. Select the first tab in the Project Navigator window and click the + sign next to **webAFCalculator** to show the components that will be selected. Two default components have already been added to the form: menuBar1 and params1. Neither of these is used for this project, but there isn't a simple way to delete them.

7. Now you are ready to start building the form visually. The first step is to insert a title for the form. Select **Text** on the Component Palette. Drag and drop a **JLabel** object onto the form in the location where you want your heading to appear.

8. Right-click on the **JLabel** control and choose **Properties**. Find the **Text** property and enter **Temperature Conversion Applet** into the text box. Change the **foreground** color to **blue** (you will get a color picker dialog box), and the **font** value (again there is a dialog box) to **Dialog-Bold-24**. Close the Properties dialog box, then right-click on the control again and select **Snap to Minimum Size**; this will size the text box to fit.

9. To create the labels for the input and results fields drag and drop another **JLabel** object onto the form. Open the Properties dialog box and type **Enter a temperature:** into the **Text** field. Change the the **font** value to **Dialog-Bold-14**. Repeat for the **Results:** label you can copy and paste the first label to save time; just remember to change the text to **Results:**.

10. Create the first input boxes by dragging and dropping a **JTextField** from the Component Palette onto the form; repeat for the second text box.

The Alignment toolbar can be used to line up the form controls, or just line them up yourself with the mouse. Note that this toolbar is enabled only when more than one control on the form is selected.

11. To add the buttons, click the **Selector** tab on the Component Palette; the **JButton** control is the first tool on the bar. Drop the control on the form and right-click on it. From the Properties dialog box, change the text to `Fahrenheit to Centigrade`. Change the font to **Dialog-Bold-14**. Close the Properties dialog box, then right-click on the control again and select **Snap to Minimum Size**.

12. Repeat for the second button, setting the text to `Centigrade to Fahrenheit`; again, copy and paste is probably the simplest strategy. The form should now look like Display B.5.

Display B.5 Completed Web Calculator Form

13. The next step is to add the formulas to the button event handlers; webAF software makes this easy. Right-click the top **JButton** and select **Handle Event**. In the New Event Handler dialog box, shown in Display B.6, under **Which type of handler do you want to create?** choose **Write your own code**.

Display B.6 New Event Handler Dialog Box

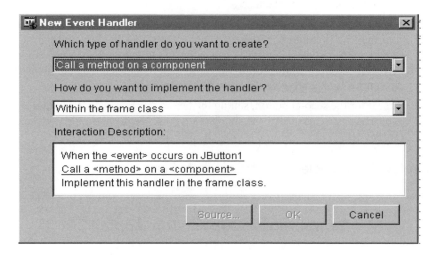

14. Click the link **When the <event> occurs on JButton1** under **Interaction Description**. The next dialog box is titled Select Source Component and Event (Display B.7). Click **actionPerformed** in the right-hand **Select the Event** pane. Click **OK**.

Display B.7 Select Source Component and Event Dialog Box

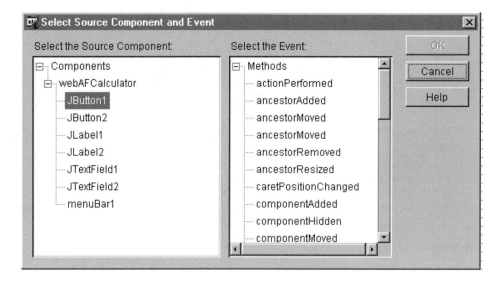

The New Event Handler dialog box should now look like Display B.8. Click **Source**.

Display B.8 Final New Event Handler Dialog Box

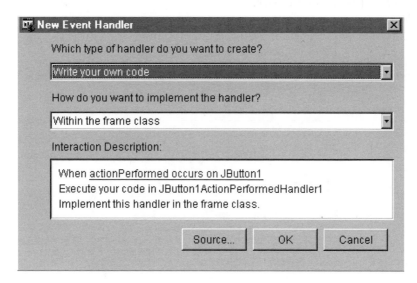

In the program code window, at the end of the Java code, you should see something like the following:

```
// JButton1ActionPerformedHandler1
// When JButton1 fires actionPerformed call
      JButton1ActionPerformedHandler1()

public void JButton1ActionPerformedHandler1
      (java.awt.event.ActionEvent event)
{
  // NOTE: Add new code here
}
```

15. After the line that says NOTE: Add new code here, enter the Fahrenheit-to-Centigrade formula:

```
double t1 = (Double.valueOf
    (JTextField1.getText())).doubleValue();
double t2 = 5*(t1-32)/9;
JTextField2.setText(String.valueOf(t2));
```

16. Click the **Visuals** tab to return to the form designer and repeat the preceding two steps for the second button. Insert the Java code for the Centigrade-to-Fahrenheit conversion:

```
double t1 = (Double.valueOf
    (JTextField1.getText())).doubleValue();
double t2 = 9*t1/5 + 32/9;
JTextField2.setText(String.valueOf(t2));
```

Actually, you'd probably want to create a subclass of JButton to do this, but let's keep it simple for the moment.

Build the project by pressing F7 (or selecting **Build** from the main menu or clicking the **Build** button on the Build toolbar).

That should do it! The resulting page (Display B.9) is displayed by clicking **Execute (Ctrl+F5)** on the webAF toolbar.

Display B.9 Temperature Conversion Applet

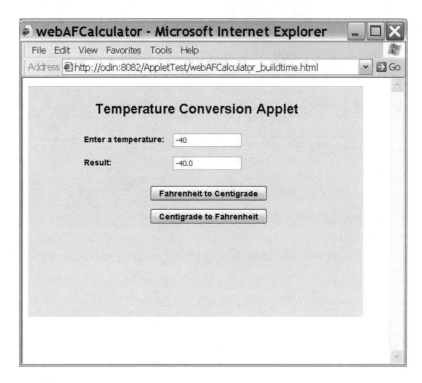

Note that webAF software has executed the command to run Internet Explorer with the following URL:

```
http://odin:8082/AppletTest/webAFCalculator_buildtime.html
```

All of the details of starting the server on port 8082 of the local host (`odin`) and locating the HTML page are taken care of for you.

The webAF Code Generator

The HTML code looks quite a bit different from Example B.1; SAS has written some fairly complex code for you:

Example B.7 webAFCalculator.html

```
<!DOCTYPE HTML PUBLIC "-//IETF//DTD HTML//EN">
<HTML>
<HEAD>
<TITLE>webAFCalculator</TITLE>
</HEAD>
<BODY>
<!{{~SAS~(APPLET) - Generated Code - Do Not Edit!>
<SCRIPT LANGUAGE="JavaScript"><!--
var _info = navigator.userAgent;
var _ns = false;
var _ie = (_info.indexOf("MSIE") > 0
```

```
        && _info.indexOf("Win") > 0
        && _info.indexOf("Windows 3.1") < 0);
//--></SCRIPT>
<COMMENT>
<SCRIPT LANGUAGE="JavaScript1.1"><!--
    var _ns = (navigator.appName.indexOf("Netscape") >= 0
             && ((_info.indexOf("Win") > 0
               && _info.indexOf("Win16") < 0
               && java.lang.System.getProperty(
"os.version").indexOf("3.5") < 0)
               || (_info.indexOf("Sun") > 0)  ||
(_info.indexOf("Linux") > 0) ));
//--></SCRIPT>
</COMMENT>
<SCRIPT LANGUAGE="JavaScript"><!--
if (_ie == true) {
  document.writeln(" <OBJECT");
  document.writeln(" CLASSID=\"clsid:8AD9C840-044E-11D1-B3E9-
00805F499D93\"");
  document.writeln("
CODEBASE=\"http://java.sun.com/products/plugin/autodl/jinstall-
1_4-windows-i586.cab#Version=1,4,0,0\"");
  document.writeln(" HEIGHT=400");
  document.writeln(" NAME=\"webAFCalculator\"");
  document.writeln(" WIDTH=600");
  document.writeln(" ID=\"webAFCalculator\"");
  document.writeln(" >");
  document.writeln(" <NOEMBED><XMP>");
}
else if (_ns == true) {
  document.writeln(" <EMBED");
  document.writeln("
PLUGINSPAGE=\"http://java.sun.com/products/plugin/index.html#
download\"");
  document.writeln(" TYPE=\"application/x-java-
applet;version=1.4\"");
  document.writeln(" HEIGHT=400");
  document.writeln(" NAME=\"webAFCalculator\"");
  document.writeln(" WIDTH=600");
  document.writeln("
ARCHIVE=\"JSASNetCopyApplet.jar,JSASNetCopy.jar\"");
  document.writeln("
CODE=\"com.sas.tools.JSASNetCopy.applet.InstallApplet.class\"");
  document.writeln("
java_codebase=\"http://hygelac/sasweb/Tools/JSASNetCopy/\"");
  document.writeln(" Applet:CODE=\"webAFCalculator.class\"");
  document.writeln(" ><NOEMBED><XMP>");
}
//--></SCRIPT>
<APPLET CODEBASE="http://hygelac/sasweb/Tools/JSASNetCopy/"
  ARCHIVE="JSASNetCopyApplet.jar,JSASNetCopy.jar" HEIGHT=400
  CODE="com.sas.tools.JSASNetCopy.applet.InstallApplet.class"
  NAME="webAFCalculator" WIDTH=600 ></XMP>
<PARAM NAME="CODEBASE"
  VALUE="http://hygelac/sasweb/Tools/JSASNetCopy/">
<PARAM NAME="ARCHIVE"
VALUE="JSASNetCopyApplet.jar,JSASNetCopy.jar">
```

```
<PARAM NAME="CODE"
  VALUE="com.sas.tools.JSASNetCopy.applet.InstallApplet.class">
<PARAM NAME="Applet:CODE" VALUE="webAFCalculator.class">
</APPLET>
</NOEMBED></EMBED></OBJECT>
<!}}~SAS~(APPLET)>
</BODY>
</HTML>
```

There are three separate JavaScripts in this program (shown in bold) which carry out the following functions:

- Create a script variable called `_ie` which is set to `true` if the user's browser is Internet Explorer (but not Windows 3.1)

- Create a script variable called `_ns` which is `true` if the user's browser is Netscape and the operating system is Windows (but not a 16-bit version of Windows or NT 3.5), Solaris, or Linux

- Write browser-dependent HTML (for Internet Explorer or Netscape) that corresponds to the generic Java Plug-in code previously described.

All of this is only necessary to provide compatibility with the two different browsers; the actual applet code follows the scripts.

The page calls the Calculator applet with four parameters:

Table B.2 Generated Applet Parameters

Parameter Name	Value	Function
ARCHIVE	JSASNetCopyApplet.jar,JSASNetCopy.jar	Java archive (JAR) files that contain the `JSASNetCopyApplet` and JSASNetCopy packages
CODE	com.sas.tools.JSASNetCopy.applet.InstallApplet.class	Specify the Java class file to run
CODEBASE	http://hygelac/sasweb/Tools/JSASNetCopy/	URL of `JSASNetCopy`
Applet:CODE	webAFCalculator.class	The applet class

The problem is that this code will only run on the local host, thanks to the CODEBASE parameter, which specifies the location as follows:

```
http://hygelac/sasweb/Tools/JSASNetCopy/
```

This location is available only from an active Web server.

The really cool part is that two separate HTML files—**webAFCalculator.html** and **webAFCalculator_buildtime.html**—are generated. There are only a few differences between the two:

The run-time HTML file contains the following code:

```
java_codebase="http://hygelac/sasweb/Tools/JSASNetCopy/");
<APPLET CODEBASE="http://hygelac/sasweb/Tools/JSASNetCopy/"
<PARAM NAME="CODEBASE"
VALUE="http://hygelac/sasweb/Tools/JSASNetCopy/">
```

The build-time file is slightly different:

```
java_codebase="http://odin:8082/sasweb/Tools/JSASNetCopy/");
<APPLET CODEBASE="http://odin:8082/sasweb/Tools/JSASNetCopy/"
<PARAM NAME="CODEBASE"
VALUE="http://odin:8082/sasweb/Tools/JSASNetCopy/">
```

This is because when webAF software was installed, the server URL was specified as http://hygelac. The IDE has made the code portable for you.

The function of JSASNetCopy is to download the required Java archive (JAR) files to the client's Java Plug-in directory if they do not already exist. To quote from the SAS documentation:

> The JSASNetCopy applet HTML (the default) is a pure Java solution that leverages the Java extension installation facilities provided by Sun Microsystems Java Plug-in software. It enables you to automatically download the most current versions of component classes that are supplied by SAS. That is, JSASNetCopy checks the dates of the appropriate JAR files for SAS classes, and then prompts you to download new JAR files if the installed JAR files for SAS classes are not the most current versions.[2]

SAS has supplied a set of additional classes that run with the Java Plug-in. These may or may not be required by a given webAF applet, so the code generator puts in the reference just in case. The important thing to note is that the JAR files must be available on the network somewhere, and the code generator has to know where they are.

The examples presented so far have not shown the most important ability of webAF software: creating dynamic, data-driven Web pages. The big advantage of SAS AppDev Studio over other Web development packages is that it provides integrated access to data as well as HTML page and Java code design tools. This power and flexibility comes at a cost, however, which is that there is a lot to learn. Once the connections and applications have been set up correctly, it should be relatively easy to develop complex data collection and retrieval applications. The hard part is just getting it all set up right. That's what the next section is all about.

Access to Remote Data with webAF Software

The following discussion assumes that you have read the section on registering connections to remote data sources in Chapter 9, "Developing Java Server-Side Applications with webAF Software."

---

[2] SAS Institute Inc., *Getting Started with AppDev Studio*, 2nd ed. (Cary, NC: SAS Institute, Inc., 2001) p.68.

Access to Remote Data

There are plenty of good Java visual development environments available. What makes webAF software particularly useful is that it includes components to automate the process of connecting to remote data sources, whether in SAS or another RDBMS such as Oracle, Sybase, or DB2. The following example shows how to accomplish this connection using Java applets and a previously defined remote connection.

1. Start by opening the webAF New Project Wizard and creating a new applet project named `TableViewExample`.

2. In the Applet Options dialog box, make sure the **Swing** radio button is selected under **Default GUI Architecture**. Click **Next** and then **Finish** to create the new applet project.

3. Make sure the **Components** tab is selected in the Project Navigator window, and you should see a blank form like the one shown in Display B.4.

4. In order to connect to a remote table, you need to add a **Data Viewer** control to the applet. Select the **Data Viewers** tab on the Component Palette. For this example, drop a **TableView** component onto the form (the second icon from the left on the **Data Viewers** component toolbar).

   Size the box by dragging on the object handles the way you would in any other Windows GUI. A new entry named **TableView1** will automatically appear on the component list in the Project Navigator window.

5. To associate this **TableView** object with the remote connection, click the **SAS** tab on the Component Palette, select the **DataSetV2Interface** control and drag it onto the **TableView** component on the form.

6. Select an interface and click **OK** in the Remote Connection dialog box; the table viewer and the interface will appear in the Project Navigator component window. The **TableView** is now associated with the **DataSetV2Interface** object.

7. The last step in getting your form to work with the remote data set is to select the table you want to display. Right-click on **dataSetV2Interface1** in the Project Navigator window and select **Customizer**. In the Customizer dialog box, click the ellipsis button next to **Data Set Name**. After a successful connection to the host, the Data Sets window will display the available libraries on the remote connection.

Choose the SASHELP library. This should result in a dialog box something like the following:

Display B.10 Remote Data Set Selection

8. Select the SASHELP.RETAIL data set as shown. Close the Customizer window.

9. Build the project by pressing F7. You now have an HTML applet that is connected to the data set on the remote system.

10. Make sure that the Java Web Server (Tomcat) is started, and then click the exclamation point on the Build-Execute toolbar. If the Java security permissions on the SAS server have been set correctly (see the following section), the table view shown in Display B.11 should appear.

You can edit this form by selecting the **Files** tab in the Project Navigator, and then right-clicking **TableViewExample.html**. Clicking **Edit** enables you to display and make changes in the HTML, such as adding a title or other tags. **Browse** opens the page in the default Web browser. Do not forget to rebuild the project after making changes to content.

Display B.11 TableView Applet (Build Time)

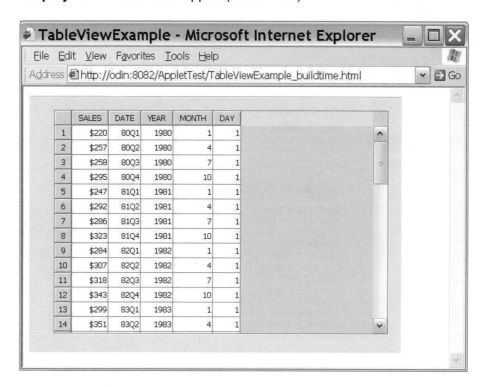

That's all there is to it—once again, SAS has done all of the work for you.

Java Security

The remote data access methods discussed in this appendix have one big disadvantage. This is related to the default security that is built in to applets. In general, Java applets run in a "sandbox"; that is, they are not allowed to access other network computers connected to the client. This provides protection from malicious downloaded code that could damage or access users' local files. Unfortunately, in order to get this example to work, it is necessary to deal with Java's built-in security systems. For testing purposes, you can modify the Java security files to allow access to remote hosts.

The Java Runtime Environment (JRE) includes a utility to change security permissions called `policytool.exe`. Running this program enables you to edit the `java.policy` file in order to change the value of the Java socket permissions. There may be several versions of this utility on your client system. Make sure you edit the correct file, which is the one that your browser uses to open applets by default.

To use the policy tool, open it from the JRE `bin` directory. You will need to navigate to the JRE `lib\security` folder, where you will find a file called `java.policy`; open it from the policy tool **File** menu. Again, make sure you have the right policy file—there may be more than one, and editing the wrong one can cause a substantial level of aggravation.

Display B.12 Java Policy Tool

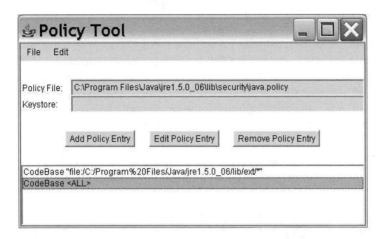

Select **CodeBase <ALL>** as shown, and then click the **Edit Policy Entry** button to set permissions. On the next screen (not shown), select the **Add Permission** button. The following dialog box should appear:

Display B.13 Add Socket Permission

As illustrated, select `SocketPermission` in the first text box. Set the target name to your server name, and choose the actions `accept, connect, listen, resolve`. In this case, specifying `hygelac:1024-` will open the client PC to all connections to port 1024 or higher on this server. Click **OK** and return to the Policy Tool window. Note that the changes will not be saved automatically; you need to select **File ▶ Save** on the Policy Tool window menu to update the `java.policy` file.

Unfortunately, it would be impractical to try to make this change on all of the client systems that will use your applets, aside from the security concerns raised by opening up this socket service. Modifying policy files is discouraged as an enterprise solution because it is difficult to automate and usually requires intervention on each client machine. Additionally, there is nothing to prevent some other application from subsequently overwriting or modifying the policy changes.

The recommended alternative is to sign and trust your applet using an RSA certificate. For more details on this, see the SAS white paper "RSA-Signing Applets," which notes the following:

> Most applets have very limited access to the computer on which they run. These intentional restrictions are designed to prevent malicious applets from taking over your computer, silently stealing your data or overwhelming your network. But what about good applets which might need to save files or talk to other computers on the network? Signing an applet can give it the access that applets are normally denied: accessing files on the client computer, connecting to arbitrary machines on the network, etc.
> (http://support.sas.com/rnd/appdev/tech/signing/RSASigning.htm)

Under the Hood: Deploying the Resulting Web Page

In order to see how to deploy the resulting Web page, open the `TableViewExample` folder in the Projects directory. Note that webAF software has created the 11 files shown in Table B.3.

Table B.3 TableViewExample Project Folder

File Name	Type	Description
TableViewExample.afx	XML	Project descriptor
TableViewExample.fml	XML	Frame descriptor
TableViewExample.wsp	Binary	Workspace file
TableViewExample.html	HTML	Web page (production)
TableViewExample_buildtime.html	HTML	Web page (local)
TableViewExample_build.xml	XML	Instructions for Ant build tool
TableViewExample.java	Java	Applet code
TableViewExample.class	Byte code	Applet class file
TableViewExample$1.class	Byte code	Applet class file
TableViewExample$2.class	Byte code	Applet class file
TableViewExample$params1InnerClass.class	Byte code	Applet class file

The first two files are in XML format and are used for project house keeping: the .afx file includes the project properties settings, while the .fml file determines the properties and the components for the frame(s). The .wsp file is automatically generated to manage the open files in the user's project workspace; it allows webAF software to remember the set of files that was open the last time the user edited the project and to re-open the same set of files on a subsequent invocation. The build XML file is used by the Apache Ant build tool to compile the project; see http://ant.apache.org/ for more information on this utility. These files are generated and read by the IDE; there is no need for the user to modify them. `TableViewExample.java` is the applet code itself; the four class files are generated when the applet is compiled.

Most of the generated HTML code consists of the JavaScripts required to load the Java Plug-in for Internet Explorer or Netscape; note the warning in **TableViewExample.html** against editing the SAS generated code. You can add HTML code below this line; be careful however, not to modify the scripts.

Using the Package Wizard to Deploy Applets

In order to run this page on a Web server, only **TableViewExample.html** and the four Java class files need to be uploaded to the Web server. The other project files are only required for development and maintenance in the webAF environment. To run the Web page on the server, simply copy these files to a public directory of the server; then type in the URL of the **TableViewExample.html** file. The applet will be loaded into the Web browser on the client and the remote data will be displayed dynamically on the client. See *Getting Started with AppDev Studio, 2nd ed.,* pp.67-69 for more detail on deploying Java applets.

As Table B.3 illustrates, webAF software generates a number of Java byte code files, which are required in order to display the applet. The webAF Package Wizard can create a single Java Archive (JAR) file containing all of the class files necessary. To use the Package Wizard, from the main webAF menu, select **Tools ▶ Wizards ▶ Package Wizard**. The initial screen is shown in Display B.14.

Display B.14 Deploying Applets: the Package Wizard

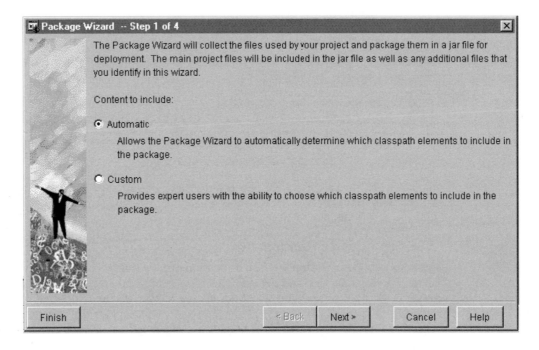

Select **Automatic** and then accept all the defaults on the remaining screens. The Package Wizard creates an **archive\jar** folder under the project directory and adds two new files there: the HTML document **TableViewExample.html**, and the JAR file **TableViewExample.jar**. Uploading these two files to a Web server HTML directory has the same effect as copying the HTML and the four class files.

In order for the applet to work correctly, it is also necessary for the following three files to exist on the server, in the **sasweb/Tools/JSASNetCopy** directory under the Web server document root. These files are installed by the SAS AppDev Studio server-side catalog updates, but it is also possible to copy them to the server manually:

- **JSASNetCopy.jar**
- **JSASNetCopyApplet.jar**
- **JSASNetCopyConfig.txt**

If the two files created by the Package Wizard have been uploaded to the public HTML directory, and the necessary server file classes exist in the correct directory, the page shown in Display B.15 should appear.

Display B.15 TableView Applet (Run Time)

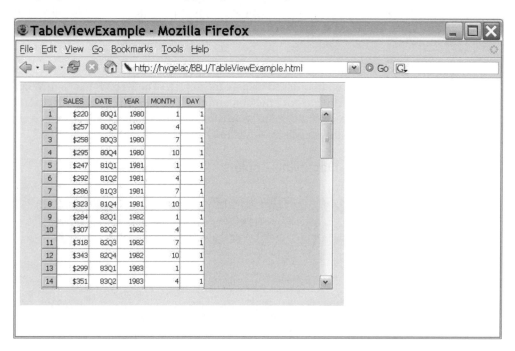

In conclusion, Java applets have a deservedly bad reputation for being slow to load. Once the Java Plug-in has been installed on the client, along with all of the archive files, applets run very quickly; getting there can take a while, however. In addition, the necessity of signing applets so that they can access other hosts has had a negative impact on the adoption of this technology. As a consequence, many sites prefer to use server-side solutions as opposed to applets. The discussion of Java servlets and JavaServer Pages in Chapter 8, "Java Servlets and JavaServer Pages," and Chapter 9, "Developing Java Server-Side Applications with WebAF Software," illustrates how they can be used to overcome the perceived problems with client-side dynamic HTML.

Appendix C

SAS/IntrNet: Design-Time Controls

Introduction

The SAS *Design-Time Controls* (DTC) are a set of ActiveX controls that can be used to create Web pages that include static or dynamic SAS content. In other words, you can display SAS files on your Web site without any Java or SAS programming. These controls (*widgets* in SAS/AF terminology) must be used from within a Microsoft Windows-based Integrated Development Environment (IDE).

Several HTML editors will work, including the following:

- Microsoft FrontPage and Visual InterDev
- Macromedia Dreamweaver
- webAF software

A description of the Design-Time Controls including the specific product versions that are supported can be found in the documentation on the "SAS Client-Side Components Volume 2" installation CD. The documentation can also be found online at http://support.sas.com/rnd/web/intrnet/dtc/.

By dropping one of these controls onto an HTML page, it is possible to provide support for users who do not have SAS installed so they can view SAS tables or graphs in a browser window on their local client machine. Although Windows is required to create the HTML pages, they can be viewed in most current Web browsers, such as Netscape Navigator or Firefox.

As noted in the documentation, there are four parts to the Design-Time Controls.

- Client-side components – the ActiveX controls themselves, contained in the file `sasdtc.ocx`.
- Dreamweaver components – optional controls for use with Dreamweaver.
- Web server components – the Application Broker, which is required for the DTC. In addition, for dynamic Web pages you must install either `AppServer.jar` for JSP or `AppServer.dll` for ASP on the server (see the following instructions).
- SAS server components – the SAS Application Server, which must be able to access `sashelp.websdk1` (see the following discussion).

Although all this sounds rather intimidating, in fact using the Design-Time Controls is one of the easiest ways to provide SAS support for HTML pages. If you are familiar with using one of the IDEs listed above, you can learn how to use the DTC in a few minutes, as the following examples illustrate. Again, reasonably clear explanations about installing all of these components are available on the SAS Web site as well as on the installation CD; please read them before calling SAS Technical Support when your programs don't work. There is also a DTC troubleshooting page at http://support.sas.com/rnd/web/intrnet/misc/support.html with solutions to many common problems that users may encounter.

Currently, seven controls are available, as shown in Table C.1.

Table C.1 SAS Design-Time Controls

Control Name	Generated Web Page Content
SAS Table	HTML table
SAS Tabular Report	HTML table from PROC TABULATE
SAS MDDB Report	Multidimensional database report in HTML form
SAS Stored Program	Output from your SAS/IntrNet program
SAS Critical Success Factor	Critical success factor Java applet
SAS Thin-Client Graphics	SAS graphic Java applet or ActiveX control
SAS Treeview	Java applet to visualize hierarchical data

Each of these controls has specific features that can be customized by the user. The following discussion focuses on the Table component, which creates an HTML table object. Using the table control example as a model, it should be possible to get a handle on how to deploy the other, more elaborate components.

Installing and Configuring the Design-Time Controls

To install the DTC, run the installation program on the "SAS Client-Side Components, Volume 2" installation CD. That's it. Since the controls are only available for Windows, there are no platform-specific installation issues.

Configuration, however, requires a little more work. The most important step is specifying the default connection—that is, the location of the Application Broker, the name of the service used, and the default folder for Java applet code. For a review of how to use the Application Broker, see Chapter 6, "SAS IntrNet: the Application Dispatcher," which covers using the SAS/IntrNet product. Although SAS DTC can be used to create ASP or Java applications the SAS Application Server is always required to provide data for the output pages, whether these have dynamic content (ASP or JSP) or are just static HTML.

Selecting **Start ▶ Programs ▶ The SAS System ▶ Design-Time Controls ▶ Configure SAS Design-Time Controls** should result in a dialog box something like Display C.1:

Display C.1 Configuring the SAS DTC

The first text box, **Perform SAS processing**, offers three options:

1. **once when building this page** – creates static HTML; the table is simply inserted into the Web page.
2. **when Java Server Page is invoked** – creates JSP to dynamically load data from the application server when the page is requested by a user.
3. **when Active Server Page is invoked** – creates dynamic ASP.

The DTC configuration utility sets the default values; you can change the connection for each individual control as necessary. On a page with multiple controls, it is actually possible (although not in MDDB) for each to be connected to a different server.

You also need to know the name of the Web server where the Application Broker is running; the assumption is that it is on the same machine as the DTC (that is, `localhost`) but you can specify another host name.

The third text box, labeled **Broker URL**, has a gotcha, at least in the DTC Version 8.2. Do not specify the full path to the broker URL; for example, do not write http://localhost/cgi-bin/broker.exe. Doing so will render the controls inoperative, although there is no way to know this from the documentation. Type the *relative* path to the Application Broker, not the *absolute* path.

The **Service** name is defined when the service is created; it is customary to create a default service called, logically enough, `default`, as shown in Display C.1. Again, use the *relative* pathname to the server; see Chapter 6 for more information about installing Application Dispatcher services.

Finally, the **Applet codebase** is `/sasweb/graph` unless you specified otherwise during installation.

Click **OK** and the default DTC configuration will be generated.

For Microsoft Internet Information Server (IIS), if you install the Web server components on the server, **AppServer.dll** will be automatically copied to the right directory and registered when you install the DTC. The Application Server library file by default is installed into `C:\Program Files\SAS\SAS Design-Time Controls`. The setup CD allows you to install just the client components, just the server components, or both. You may need to ask your system administrator to install the required components on the server.

As noted above, in order to provide dynamic content using JavaServer Pages the file **AppServer.jar** must be available on the Web server. For Tomcat 4.0, copy the file from the installation directory on the local host (`C:\Program Files\SAS\SAS Design-Time Controls`) to the **server/lib** subdirectory under the Tomcat root.

Open the Application Broker startup program **appstart.sas** in the service directory on the Web server. This SAS program file should include the following line somewhere following the PROC APPSERV statement:

```
proglibs sashelp.websdk1;
```

Add it if it is not already there. Now restart the application server and you should be in business.

Adding a SAS Table to an HTML Page

To create a table, start with a new blank HTML page. The procedures for Dreamweaver and webAF software are generally similar; examples using the two products follow the more detailed FrontPage 2003 example.

Microsoft FrontPage

To create a table in Microsoft FrontPage 2003, Select **Insert ▶ Web Component ▶ Advanced Controls ▶ Design Time Control** from the main FrontPage 2003 menu and click the **Next** button as shown in Display C.2.

Display C.2 Microsoft FrontPage 2003: Insert Web Component

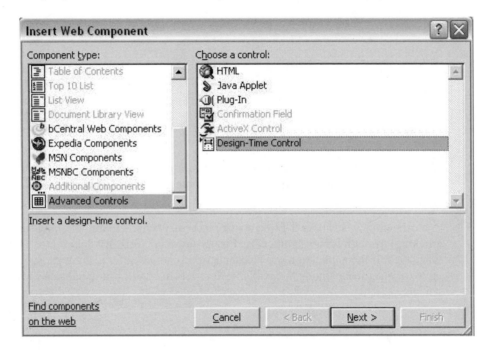

The first time you do this, the next dialog box will display an empty list box.

Select the **Customize** check box and click the specific controls that you would like to include. Choosing **Select All** and **Finish** results in the window shown in Display C.3:

Display C.3 Microsoft FrontPage 2003: SAS Design-Time Controls

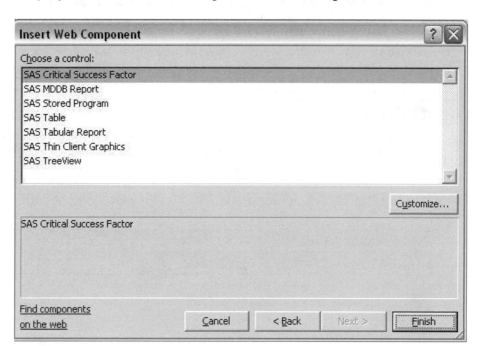

For this example, choose **SAS Table** and then **Finish**. Depending on your Windows settings, you may get an ActiveX control warning. If so click **Yes**; you do want to run the control.

You should now see a SAS Table control icon displayed. Double-clicking (or right-clicking) the icon brings up the DTC Properties dialog box. By default, the **Dataset** tab is displayed. If the application server connection is active, after a slight pause you will be able to select from among the available data sets in the drop-down list as shown in Display C.4.

Display C.4 SAS Design-Time Controls: Select Data Set

If no data sets show up, you need to go to the **Connection** tab in the DTC Properties dialog box to make sure that you have correctly specified the connection. You should have access to all of the libraries that are allocated by the `appstart.sas` program (see Chapter 6 for a discussion of allocating libraries). Make sure the application server is running; if you installed it as a service, try restarting the service from **Start ▶ Programs ▶ The SAS System ▶ IntrNet ▶ default Service ▶ Start Windows Service**.

After modifying or changing connections, you will need to close the DTC Properties dialog box and reopen it for the data set drop-down list to repopulate; the list will not show up automatically after clicking **Apply** on the **Connection** tab.

Back at the **Dataset** tab, you can also specify a query to be executed against the selected data, as shown in Display C.5.

Display C.5 SAS Design-Time Controls: Select Observations

In this case, only records for 1994 will be displayed.

Add title and a couple of header lines to decorate the page and save it into your local Web server directory or copy it to a remote Web server. Display C.6 shows the result.

Display C.6 SAS Design-Time Controls: Sample Table Output

Note that this is a static page. It will contain this data even if the underlying SAS data set changes. Viewing the page source, shown in Example C.1, demonstrates that the data was inserted into the page by the ActiveX control when it was added to the page.

As is characteristic of visual HTML tools, the resulting code ain't pretty, but it works. Note that the code generated by DTC should never be modified by the user; DTC "owns" the HTML and will overwrite any modifications you may make. Consequently, you should not make any changes to the code between the two metadata statements shown in bold in the example.

Example C.1 Generated HTML Source Code

```
<html>
<head>
  <title>SAS Design Time Controls Example</title>
</head>
<body>
<h1>DTC Table Control: SASHELP.RETAIL</h1>
<p><!--metadata type="DesignerControl" startspan
<object classid="clsid:6ED5010A-D596-11D3-87D7-00C04F2C0BF6"
id="Table1" dtcid="2">
  <param name="BrokerURL" value="/cgi-bin/broker">
  <param name="DataSet" value="SASHELP.RETAIL">
  <param name="Debug" value="0">
  <param name="Password" value>
  <param name="ProcessingMode" value="build">
  <param name="Program" value="sashelp.websdk1.ds2htm2.scl">
  <param name="Service" value="default">
  <param name="SQLView" value>
  <param name="WebServer" value="hygelac">
  <param name="WhereClause" value="year eq 1994">
  <param name="PagePart" value="body">
  <param name="DisplayVariables" value>
  <param name="IDVariables" value>
  <param name="ByVariables" value>
  <param name="SumVariables" value>
  <param name="TranscodingList" value>
  <param name="PropertiesList" value>
  <param name="Encode" value>
  <param name="CharacterSet" value>
  <param name="TableWidth" value="0">
  <param name="TableWidthUnits" value="Percent">
  <param name="Border" value="Y">
  <param name="BorderWidth" value>
  <param name="TableBackgroundColor" value>
  <param name="ObsBackgroundColor" value>
  <param name="IDVariableBackgroundColor" value>
  <param name="VariableBackgroundColor" value>
  <param name="SumVariableBackgroundColor" value>
  <param name="ColumnLabelBackgroundColor" value>
  <param name="TableCaption" value>
  <param name="NumericRoundOff" value>
  <param name="CellPadding" value>
  <param name="CellSpacing" value>
  <param name="Formats" value="Y">
  <param name="Labels" value="Y">
  <param name="DisplayObs" value="N">
  <param name="TableAlign" value="default">
  <param name="VariableForegroundColor" value>
  <param name="SumVariableForegroundColor" value>
  <param name="ColumnLabelForegroundColor" value>
  <param name="IDVariableForegroundColor" value>
```

```
    <param name="ObsForegroundColor" value>
    <param name="ByVariableForegroundColor" value>
    <param name="VariableHorizontalAlign" value>
    <param name="SumVariableHorizontalAlign" value>
    <param name="ColumnLabelHorizontalAlign" value>
    <param name="IDVariableHorizontalAlign" value>
    <param name="ObsHorizontalAlign" value>
    <param name="CaptionHorizontalAlign" value>
    <param name="VariableVerticalAlign" value>
    <param name="SumVariableVerticalAlign" value>
    <param name="ColumnLabelVerticalAlign" value>
    <param name="IDVariableVerticalAlign" value>
    <param name="ObsVerticalAlign" value>
    <param name="CaptionVerticalAlign" value>
    <param name="VariableFontFace" value="Arial,Helvetica,sans-
serif">
    <param name="SumVariableFontFace"
value="Arial,Helvetica,sans-serif">
    <param name="ColumnLabelFontFace"
value="Arial,Helvetica,sans-serif">
    <param name="IDVariableFontFace" value="Arial,Helvetica,sans-
serif">
    <param name="ObsFontFace" value="Arial,Helvetica,sans-serif">
    <param name="CaptionFontFace" value="Arial,Helvetica,sans-
serif">
    <param name="ByVariableFontFace" value="Arial,Helvetica,sans-
serif">
    <param name="ExtraParms" value>
    <param name="RegistryKey"
    value="Software\SAS Institute Inc.\SAS Design-Time
Controls\TableCtrl">
</object>
-->

<P>

<TABLE BORDER="1" WIDTH="0%" CELLPADDING="1" CELLSPACING="1">
  <TR>
    <TH ALIGN="CENTER" VALIGN="MIDDLE"><FONT
FACE="Arial,Helvetica,sans-serif"> Retail sales in millions of
$</FONT></TH>
    <TH ALIGN="CENTER" VALIGN="MIDDLE"><FONT
FACE="Arial,Helvetica,sans-serif"> DATE</FONT></TH>
    <TH ALIGN="CENTER" VALIGN="MIDDLE"><FONT
FACE="Arial,Helvetica,sans-serif"> YEAR</FONT></TH>
    <TH ALIGN="CENTER" VALIGN="MIDDLE"><FONT
FACE="Arial,Helvetica,sans-serif"> MONTH</FONT></TH>
    <TH ALIGN="CENTER" VALIGN="MIDDLE"><FONT
FACE="Arial,Helvetica,sans-serif"> DAY</FONT></TH>
  </TR>
```

```
<TR>
    <TD ALIGN="CENTER" VALIGN="MIDDLE"><FONT
FACE="Arial,Helvetica,sans-serif">      $876</FONT></TD>
    <TD ALIGN="CENTER" VALIGN="MIDDLE"><FONT
FACE="Arial,Helvetica,sans-serif">9 4Q1</FONT></TD>
    <TD ALIGN="CENTER" VALIGN="MIDDLE"><FONT
FACE="Arial,Helvetica,sans-serif"> 1994</FONT></TD>
    <TD ALIGN="CENTER" VALIGN="MIDDLE"><FONT
FACE="Arial,Helvetica,sans-serif"> 1</FONT></TD>
    <TD ALIGN="CENTER" VALIGN="MIDDLE"><FONT
FACE="Arial,Helvetica,sans-serif"> 1</FONT></TD>
  </TR>
  <TR>
    <TD ALIGN="CENTER" VALIGN="MIDDLE"><FONT
FACE="Arial,Helvetica,sans-serif">      $998</FONT></TD>
    <TD ALIGN="CENTER" VALIGN="MIDDLE"><FONT
FACE="Arial,Helvetica,sans-serif"> 94Q2</FONT></TD>
    <TD ALIGN="CENTER" VALIGN="MIDDLE"><FONT
FACE="Arial,Helvetica,sans-serif"> 1994</FONT></TD>
    <TD ALIGN="CENTER" VALIGN="MIDDLE"><FONT
FACE="Arial,Helvetica,sans-serif"> 4</FONT></TD>
    <TD ALIGN="CENTER" VALIGN="MIDDLE"><FONT
FACE="Arial,Helvetica,sans-serif"> 1</FONT></TD>
  </TR>
</TABLE>
<P>

<!--metadata type="DesignerControl" endspan--></p>

</body>

</html>
```

Macromedia Dreamweaver

In Macromedia Dreamweaver 2 and later, the process of inserting the table is even easier than in FrontPage.[1] Just select **Insert** from the main menu and the controls appear as shown in Display C.7.

From that point on, the process of customizing the page is as previously described.

---

[1] Macromedia's Dreamweaver MX HTML editor is not supported for SAS Design-Time Controls (see SAS Note 010249).

Display C.7 Inserting SAS DTC in Dreamweaver 3

webAF Software

webAF software can also be used to insert controls. Open a new or existing project and insert a new HTML page. (Select **File ▶ New** or press CTRL+n). The DTCs are available from the **Insert** menu as shown in Display C.8.

Display C.8 Inserting SAS DTC in webAF Software

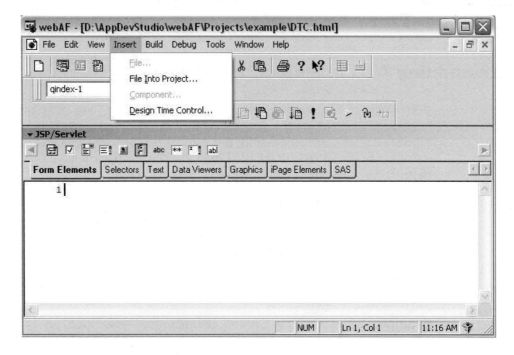

Selecting **Design Time Control** brings up a pop-up dialog box:

Display C.9 Insert DTC Dialog Box

Select a control and proceed as previously described for FrontPage to customize it.

Whichever HTML editor you use, the result is the same: the ActiveX control inserts a static table into the page. If you want the table to update dynamically, you can also use DTC to create JSP or ASP pages.

Adding Dynamic SAS Content with the Design-Time Controls

Embedding DTC in a JavaServer Page

To create a JSP to display a dynamic SAS table, create a blank HTML page and insert a Table control as described in the preceding section. On the **Connection** tab of the DTC Properties dialog box, make sure that **Perform SAS processing when Java Server Page is invoked** is selected in the first list box. Do this first, before selecting the data set to use. Then, select **SASHELP.RETAIL** from the **Dataset** tab and, if you wish, a select statement as shown above. Click **Apply** to insert the control.

The resulting HTML page contains the JSP scriptlet shown as Example C.2 (the Java code has been reformatted for legibility). Note that the URL property of the `AppServer` object is set to point to the Application Broker by the following statement:

```
IappServer.setURL("http://hrothgar/cgi-bin/broker");
```

This is why SAS/IntrNet is required even for ASP and JSP using the DTC.

Example C.2 DTC Table Control: JavaServer Page Code

```
<%
{
  appserver.AppServer IappServer = new appserver.AppServer();
  IappServer.setURL(null);
  IappServer.setURL("http://hygelac/cgi-bin/broker");

  String queryString =
"_service=default&_debug=0&_program=sashelp.websdk1.ds2htm2.scl
&data=SASHELP.RETAIL&pagepart=body&sqlview=Y&where=year%20eq%20
1994&twidth=0&twunits=Percent&border=Y&formats=Y&labels=Y&obsnum=
N&talign=default&vface=Arial,Helvetica,sans-
serif&sface=Arial,Helvetica,sans-
serif&clface=Arial,Helvetica,sans-
serif&iface=Arial,Helvetica,sans-
serif&oface=Arial,Helvetica,sans-
serif&cface=Arial,Helvetica,sans-
serif&bface=Arial,Helvetica,sans-serif";

  java.io.OutputStream os = IappServer.getOutputStream();
  os.write(queryString.getBytes());
  os.close();

  String HTML = IappServer.getHTML();

  int responseCode = IappServer.getResponseCode();

  if (responseCode >= 200 && responseCode < 300)
  out.println(HTML);
  else if (responseCode == 401) {
        HTML = "<P><FONT COLOR=\"red\">ERROR: Authenticated
        sites are notsupported.</FONT></P><BR>" + HTML;
        IappServer.printDTCError(new java.io.PrintWriter(out,
```

```
true),
      responseCode, HTML);
   }
 else
  IappServer.printDTCError(new java.io.PrintWriter(out, true),
      responseCode, HTML);
}
%>
```

The scriptlet just creates a new `AppServer` object and sends it a query. If the response code from the application server is in the range 200–299, the table is displayed; otherwise you get an error message.

Save the resulting page with a .jsp extension and move it to the Tomcat **webapps** directory (see Chapter 8, "Java Servlets and JavaServer Pages," for a discussion of deploying JSP). Open the page in your favorite browser, and you should see something that looks very similar to Display C.6.

Display C.10 SAS DTC: JavaServer Page Output

The only difference between this display and the one shown in Display C.6 is that this table is dynamic. Changing the values in the underlying SAS data set will result in different values being displayed to the user.

Embedding DTC in an Active Server Page

Creating an Active Server Page works the same way. Select **Perform SAS processing when Active Server Page** is invoked, and the following script is generated.

Example C.3 SAS DTC: Table Control: Active Server Page Code

```
<%
sub displayURL(URL)
  Dim AppServer, HTML

  Set AppServer = CreateObject("SAS.AppServerPostURL")

  AppServer.webServer = "hunding"
  AppServer.URL = "/scripts/broker.exe"
  AppServer.queryString = URL
```

```
        HTML = AppServer.openURL()

    Response.Write HTML
End sub

displayURL("_service=default&_debug=0&_program=sashelp.websdk1.
ds2htm2.scl&data=SASHELP.RETAIL&pagepart=body&twidth=0&twunits=
Percent&border=Y&formats=Y&labels=Y&obsnum=N&talign=default&vfa
ce=Arial,Helvetica,sans-serif&sface=Arial,Helvetica,sans-
serif&clface=Arial,Helvetica,sans-
serif&iface=Arial,Helvetica,sans-
serif&oface=Arial,Helvetica,sans-
serif&cface=Arial,Helvetica,sans-
serif&bface=Arial,Helvetica,sans-serif")
%>
```

Note the similarity to the JSP shown in Example C.2. As with the Java code, the `AppServer` URL is set to point to the Application Broker, so you must have SAS/IntrNet installed in order to make use of the DTC.

Save the page with an `asp` extension into the desired Web folder—for example, `C:\inetpub\wwwroot\sasweb`. If your Web server is not your client machine, then you will also need to copy the file **AppServer.dll** to the .. folder on the server.

You should then be able to view the resulting dynamic Web page using IIS as the Web server.

References

Links

- HTML Editors –
  http://dir.yahoo.com/Computers_and_Internet/Software/Internet/
  World_Wide_Web/HTML_Editors/

- Microsoft FrontPage - http://www.microsoft.com/frontpage/

- Macromedia Dreamweaver –
  http://www.macromedia.com/software/dreamweaver/

- Requirements for the SAS Design-Time Controls –
  http://support.sas.com/rnd/web/dtc/require.html

- What Are SAS Design-Time Controls? –
  http://support.sas.com/rnd/web/dtc/overview.html

Appendix D

Online Analytic Processing with webEIS Software

Introduction

The easiest way to create dynamic online documents with SAS AppDev Studio is by using webEIS software. If webAF software is the Swiss army knife of Web development then webEIS software is more like the sushi chef's precision blade, specialized to slice and dice data in a variety of interesting ways.[1] The major advantage of webEIS software is that it is a completely visual development environment; no coding is required. Although webEIS software is used to generate Java applets and JavaServer Pages, no

---

[1] According to the staff at SAS, webEIS is currently in maintenance-only mode, because it is being phased out.

Java programming experience is necessary. As the webEIS Help documentation points out:

> To create a document, simply drag and drop the components that you want to use (charts, tables, images), choose a database, and then specify how you want the data organized. Finally, save your document as an applet or JavaServer Page.

The disadvantage, as with any specialized tool, is that webEIS software does only one thing—analyze SAS data in SAS MDDB (multidimensional database) form—although it does that well.

The available display formats include tables, descriptive statistics, and a variety of graphs, but only a limited set of analytic capabilities is provided. In addition, only SAS data can be analyzed—there is no way to link to Oracle, for example, as one can using JDBC with webAF software. Consequently, webEIS software will be most useful to SAS users who are already familiar with the SAS MDDB server or the SAS OLAP Server, which has replaced the earlier product server in SAS®9 (see http://www.sas.com/technologies/dw/storage/mddb/ for more information about SAS multidimensional database software).[2]

webEIS software differs from more traditional data analysis in another way as well. Long-time SAS users appreciate the flexibility of SAS software. MDDB, in contrast, depends on a predefined structure; if the database contains a particular tabulation, MDDB can generate it with remarkable speed and efficiency. If it does not, you are stuck. Of course that only means that you need to turn to some of the older SAS analytic tools, which complement webEIS with powerful and sophisticated data modeling components.

An additional limitation is that webEIS Software is a Java application that requires the Java Runtime Environment (JRE) 1.4.1, which is no longer supported by Sun. You can still download it from http://java.sun.com/products/archive/j2se/1.4.1_07/, but you are warned that the use of this product "is at the sole discretion of the end user and Sun assumes no responsibility for any resulting problems."

Given the constraints noted above, webEIS software provides astonishingly powerful capabilities for OLAP. Once the MDDB structures have been created, developers can easily create dynamic Web pages to access the data, and users can readily utilize the results to produce a variety of complex reports.

Creating a webEIS Document

To create a new webEIS application, open the design interface by selecting it from the Windows **Start ▶ SAS AppDev Studio ▶ webEIS** menu. webEIS uses a *document* model; a single document can contain multiple *sections*. Each document is stored as a binary file with a file extension of .eis, and corresponds to a collection of Web pages. Every document must have at least one section. Consequently, the initial window for webEIS looks something like the following:

---

[2] SAS does offer a one-day course called "SAS Web Tools: Accessing MDDB Data Using webEIS Software," but it seems to be available only in Australia and New Zealand; see http://www.sas.com/offices/asiapacific/sp/training/courses/weis2.html.

Display D.1 Create a New Document

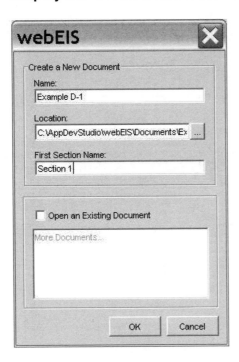

You can create a new document or open an existing one. As illustrated, the default directory is the `webEIS\Documents` folder in the SAS AppDev Studio installation directory. To create a new document, you must supply both a document name and the name of the first section.

Creating a webEIS Connection

The next window, shown in Display D.2, is used to select the data source for the first section. Each section is linked to a single data source, but multiple sections can easily use the same data source.

A webEIS data source is a SAS MDDB accessed via the SAS/CONNECT spawner (see Chapter 4, "Remote Access to SAS") or the IOM spawner (see Chapter 9, "Developing Java Server-Side Applications with webAF Software"). For a given document, you can select an existing document connection or create a new one. A webEIS document connection is also a webAF *connection* object, but it is important to note that not all webAF connections are data sources. In order to make a webAF connection available to webEIS software, it must be *imported*—that is, attached to the document.

Initially, there are no server connections defined to webEIS software. To create a connection, click on the ellipsis button next to the **Server Connection** list box. Select **New** in the pop-up Document Connections window. Fill out the information in the Connection Manager window. The format is the same as for a webAF connection, as described in Chapter 9.

Display D.2 Connect to a webEIS Data Source

Three types of data sources are available in the list box:

- Multidimensional Database (MDDB) data
- EIS Application data
- Section Link, which re-uses the data source from a previous section

In Display D.2, the first option has been selected; webEIS software will then display only the data available of that type on the server. By default, the only MDDB sources shown are in the SASHELP library. In order to access other MDDB files, the data repository must be registered. See the SAS Technical Support Notes for more information about MDDB repositories; see SN-003835, "Accessing Your Repository Registrations from within AppDev Studio and webAF or webEIS" at http://support.sas.com/techsup/unotes/SN/003/003835.html.

For this example, the PRDSALE table has been selected. The remaining tabs shown are optional—to select this data source, just click **OK**. This will opens webEIS software, as shown in Display D.3.

Creating a webEIS Table

Like Caesar's Gaul, the webEIS main window is divided into three parts. At the top of the window are the file menu and the standard and formatting toolbars. There are also several document run-time controls placed at the bottom.

The Navigation window on the left includes three tabs:

- Contents – a list of document sections
- Data – the contents of the document data source
- Bookmarks – for saving specific views of particular analyses (see the next section)

In the example, the **Data** tab is selected. The upper portion of the Data pane contains the webEIS Query list, representing the underlying database query. Initially, the window will show only stubs that can be used to view data depending on the model selected.

The rest of the pane is the MDDB list—that is, the data from the selected table or MDDB file. The PRDSALE table includes 10 variables. In the MDDB list shown, these are organized into three sets of classification variables—**Geographic**, **Product Line**, and **Time**—and two analysis variables—**Actual Sales** and **Predicted Sales**. Two additional computed variables are included in the MDDB file—**Sales Lag** and **Sales Ratio**.

The large window on the right is the Design/Preview window; the two modes toggle back and forth. The **View** menu provides a visible indicator as to which is selected. (Also, the toggle buttons at the top are an indicator; while in preview mode, the design mode button is dimmed, and vice versa.) In the Design view you can format the tables or graphs you want, whereas the Preview mode supports drilldown and the other navigation actions.

Display D.3 webEIS Design Interface

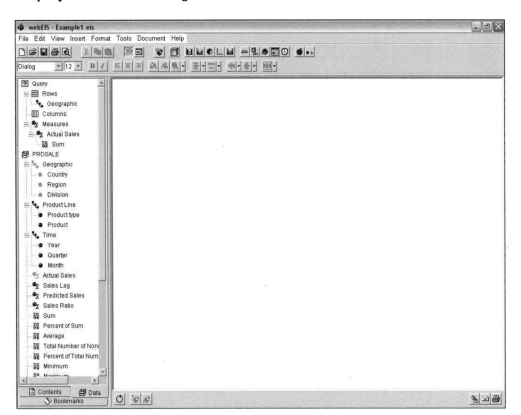

The Quick Start menu in webEIS Help provides an example of how to use the Design view to create a table. The following sequence of steps should work:

To create a table, start by dragging a Table widget (it looks like a cube) from the webEIS standard toolbar onto the Design window area to the right of the Navigation pane.

Right-click on the **Rows** entry in the Navigation pane. Two choices are available: **Select Levels** and **Refresh**. Selecting the first of these brings up the following dialog box:

Display D.4 Select Table Rows

For this example, select **Geographic** as shown and click **OK**.

Clicking a classification variable will create a new row; in order to include multiple variables, hold down the CTRL key and click on all of the items desired. (The order you click is the order they will appear in the table.) Note that the analytic variables do not appear—they cannot be selected as row-level variables. The row selection dialog box is a *modal* dialog box; you must choose **OK** or **Cancel**; click **OK** to confirm the selections before continuing.

Repeat the process to insert columns; the **Columns** item is below **Rows** in the Query list. In this case, select **Product Line**.

Right-clicking the **Measures** item beneath **Columns** provides three choices:

- **Select Analysis Variables**
- **Select Statistics**
- **Refresh**

Display D.5 shows the list of analysis variables from the first option; again, holding down the CTRL key while clicking **Actual Sales** and **Predicted Sales** includes those variables in the order selected.

Display D.5 Select Analysis Variables

The analytic variable will be displayed under **Measures** in the Navigation window.
Right-click on a variable name to select the statistics desired. Display D.6 shows the
statistics available from the second selection. A total of 19 descriptive statistics are
available, ranging from Sum to Probability of a Greater Absolute Value. This is basically
the set available in the SUMMARY, TABULATE and REPORT procedures, which also
provide cross-classification capabilities (but not dynamically).

Display D.6 Select Statistics

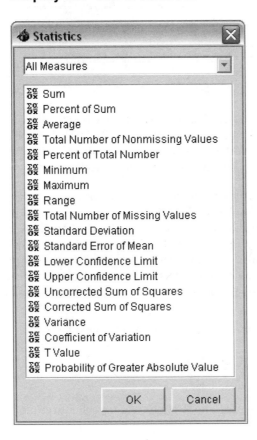

Note that the row variable will not appear until the table is refreshed. There are at least three ways to do this:

- Choose **View ▶ Refresh** from the main menu.
- Right-click **Rows** in the Query list and select **Refresh**.
- Click the **Refresh** button below the lower left-hand corner of the Design pane.

Whichever one you choose, webEIS software will retrieve the current values for the selected hierarchy from the database connection.

Selecting **Geographic** as the row classification, **Product Line** for the columns, and the analysis variables **Actual Sales** and **Predicted Sales** from **Measures**, along with **Sum** from the statistics menu should generate the table shown in Display D.7 (do not forget to refresh the table in the Design pane).

Display D.7 Sample Table Output

To demonstrate the classification capabilities of MDDB, switch to the Preview pane by selecting **View ▶ Preview Mode** from the webEIS menu. Right-clicking on a country name brings up another dialog box, shown in Display D.8, with a variety of analysis options. These options are context sensitive, and include the following:

- Drilldown/Up – displays the data at the next or previous level of the hierarchy. The difference between **Down** and **Drilldown** is that **Down** shows the next level of values for ALL the members at the current level, while **Drilldown** shows the next level of values for ONLY the member that you clicked on. For example, if you have `Time` on the rows, and you click on 1993 and choose **Down**, you will see the `Quarter` information for all years in the table. However, if you click on 1993 and choose **Drilldown**, you will see only the `Quarter` information for 1993.
- Expand/Collapse – toggles the display of data at the next level while keeping the current level visible.
- Show Detail Data – displays the underlying observations.
- Subsets – displays only selected data.
- Exception Highlighting – uses a different format for cell ranges meeting specified conditions.
- Export to Excel – displays selected data as a spreadsheet.

The number of options is quite large, and which choices are displayed is context-sensitive; that is, it depends on where you are in the hierarchy. The best way to figure out what all these do is just to experiment with the sample data provided.

Display D.8 Table Display Options

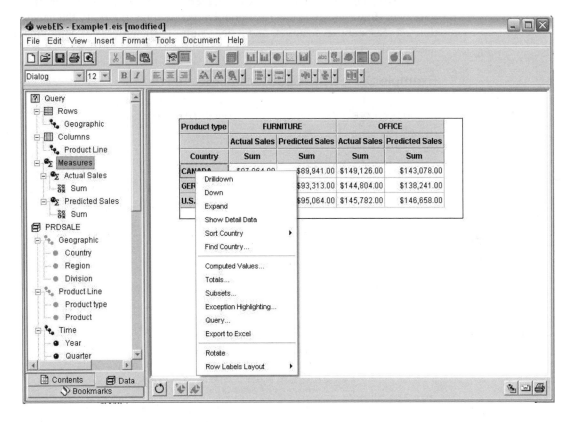

Displaying webEIS Documents

The default output mode for documents is as Java applets; it is also possible to save them as JavaServer Pages. The difference, as noted previously, is that applets are displayed on the client and require downloading a Java browser plug-in, while JSP runs on the Web server and is not usually sensitive to which Web browser is used.

Selecting **File ▶ Save** or **Save As** from the webEIS menu just saves the document with an extension of .eis, as noted above. However the main **File** menu also includes three additional choices: **Preview in Browser**, **Save as Applet** and **Save as JSP**. The first two of these do the same thing; that is, they create a Java applet that can be loaded into a Web browser (assuming the correct plug-in is available). Previewing a document creates a Web page called **_webEIS_preview_.html** that calls an applet contained in the Java archive file **_webEIS_preview_.jar**.[3]

---

[3] You also get a file called **_webEIS_preview_.eis** which can also be loaded into webEIS.

Applets

Choosing **Save as Applet** from the **File** menu generates the HTML page shown in Example D.1; comparing this code with the pages generated by webAF software displays the common family lineage. Note that since the included Java applet code (as opposed to the HTML page) is created as a JAR file, there is no way to look at it in the webEIS interface. However, since the user should never attempt to modify the generated code, this is a reasonable precaution, and it does simplify the process of deploying the resulting applet code.

Example D.1 HTML Code for webEIS Applet

```
<!DOCTYPE HTML PUBLIC "-//IETF//DTD HTML//EN">
<HTML>
<HEAD>
<TITLE></TITLE>
</HEAD>
<BODY>
<!{{~SAS~(APPLET) - Generated Code - Do Not Edit!>
<SCRIPT LANGUAGE="JavaScript"><!--
var _info = navigator.userAgent;
var _ns = false;
var _ie = (_info.indexOf("MSIE") > 0
     && _info.indexOf("Win") > 0
     && _info.indexOf("Windows 3.1") < 0);
//--></SCRIPT>
<COMMENT>
<SCRIPT LANGUAGE="JavaScript1.1"><!--
    var _ns = (navigator.appName.indexOf("Netscape") >= 0
             && ((_info.indexOf("Win") > 0
         && _info.indexOf("Win16") < 0
         &&
java.lang.System.getProperty("os.version").indexOf("3.5")<0)
         || (_info.indexOf("Sun") > 0)  ||
(_info.indexOf("Linux")>0) ));
//--></SCRIPT>
</COMMENT>
<SCRIPT LANGUAGE="JavaScript"><!--
if (_ie == true) {
  document.writeln(" <OBJECT");
  document.writeln(" CLASSID=\"clsid:8AD9C840-044E-11D1-B3E9-
     00805F499D93\"");
  document.writeln("
  CODEBASE=\"http://java.sun.com/products/plugin/1.3.0_01/
     jinstal l-130_01-win32.cab#Version=1,3,0,1\"");
  document.writeln(" HEIGHT=790");
  document.writeln(" WIDTH=1026");
  document.writeln(" >");
  document.writeln(" <NOEMBED><XMP>");
}
else if (_ns == true) {
  document.writeln(" <EMBED");
  document.writeln("
     PLUGINSPAGE=\"http://java.sun.com/products/plugin/
     1.3.0_01/plugin-install.html\"");
  document.writeln(" TYPE=\"application/x-java-
     applet;version=1.4\"");
```

```
    document.writeln(" HEIGHT=790");
    document.writeln(" WIDTH=1026");
    document.writeln("
       ARCHIVE=\"JSASNetCopyApplet.jar,JSASNetCopy.jar\"");
    document.writeln("
       CODE=\"com.sas.tools.JSASNetCopy.applet.InstallApplet.
       class\"");
    document.writeln("
       java_codebase=\"http://localhost/sasweb/Tools/
       JSASNetCopy/\"");
    document.writeln(" reportURL=\"Example D-1.eis\"");
    document.writeln("
       Applet:CODEBASE=\"http://localhost:8082/sasweb/Tools/
       JSASNetCopy\"");
    document.writeln("
       Applet:CODE=\"com.sas.tools.JSASNetCopy.applet.Install
       Applet.class\"");
    document.writeln("
       Applet:ARCHIVE=\"JSASNetCopyApplet.jar,JSASNetCopy.
       jar\"");
    document.writeln(" ><NOEMBED><XMP>");
}
//--></SCRIPT>
<APPLET CODEBASE="http://localhost/sasweb/Tools/JSASNetCopy/"
ARCHIVE="JSASNetCopyApplet.jar,JSASNetCopy.jar" HEIGHT=790
CODE="com.sas.tools.JSASNetCopy.applet.InstallApplet.class"
WIDTH=1026 ></XMP>
<PARAM NAME="CODEBASE"
VALUE="http://localhost/sasweb/Tools/JSASNetCopy/">
<PARAM NAME="ARCHIVE"
VALUE="JSASNetCopyApplet.jar,JSASNetCopy.jar">
<PARAM NAME="CODE"
VALUE="com.sas.tools.JSASNetCopy.applet.InstallApplet.class">
<PARAM NAME="reportURL" VALUE="Example D-1.eis">
<PARAM NAME="Applet:CODEBASE"
VALUE="http://localhost:8082/sasweb/Tools/JSASNetCopy">
<PARAM NAME="Applet:CODE"
VALUE="com.sas.tools.JSASNetCopy.applet.InstallApplet.class">
<PARAM NAME="Applet:ARCHIVE"
VALUE="JSASNetCopyApplet.jar,JSASNetCopy.jar">
</APPLET>
</NOEMBED></EMBED></OBJECT>
<!}}~SAS~(APPLET)>
</BODY>
</HTML>
```

By default, the HTML and JAR files will be saved to the document directory specified when SAS AppDev Studio is installed.

Unfortunately, in the move from SAS AppDev Studio 2.0 to SAS AppDev Studio 3.0, webEIS software lost the ability to display applets remotely. You can certainly display this document from within webEIS software, but if you need to move the document to a different platform, you must use JSP or ASP.

JavaServer Pages

If the webEIS document is to be shared on the World Wide Web, a good choice might be to save the document as a JavaServer Page. Selecting **File ▶ Save as JSP** saves the document as a set of JavaServer Pages to the specified directory; it is also possible to zip the required files by checking the corresponding box in the Save File dialog box. In this way, only one file need be copied to the **$TOMCAT_HOME/webapps** directory on the Web server. Unzip the package and then stop and restart Tomcat.

It is important to note that in order to run webEIS JSP on a Web server, the SAS AppDev Studio JAR files should be copied to the **$JRE_HOME/lib** directory on the server. (See the discussion of deploying JSP in Chapter 9.) The files are located in the **\AppdevStudio\java\sas\ext** directory on the Windows system.

Display D.9 shows the resulting page in a browser window.

Display D.9 JSP Output

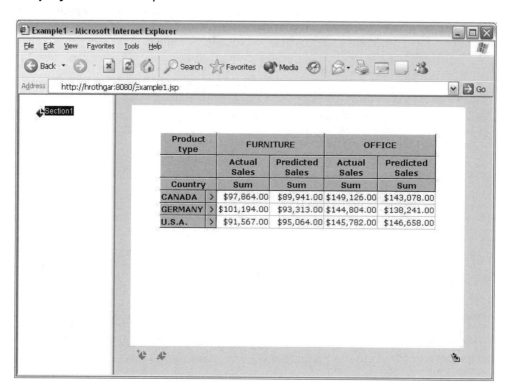

The URL `http://Hrothgar:8080/Example1.jsp` opens the page in Tomcat, displaying the same information as the applet shown in Display D.7. The big difference is that using the Web server to generate HTML code requires a round trip over the TCP/IP connection plus an additional round trip connection between the Web server and the SAS/CONNECT server. If the servers are slow or heavily burdened, the resulting delays can be quite painful.

The other major difference between accessing a webEIS applet and JSP is that with JSP, right-clicking on an item does not bring up the menu of choices shown in Display D.8. Instead, as can be seen from the preceding example, you need to click on a directional arrow in order to display additional information or move back up the hierarchy.

In general, webEIS documents work better on the client than on the server. This is an exception to the rule that you should use the server to provide Web content whenever possible. Unless your server is very fast and relatively lightly utilized, you should use the webEIS client where possible.

Additional webEIS Controls

In addition to multidimensional tables, webEIS software offers five kinds of graphical displays:

- bar charts
- segmented bar charts
- pie charts
- scatter plots
- combination charts

There is a button for each on the standard tool bar. Charts provide many of the same features as tables, including drilldown, subsets, and exporting to Excel. Charts are constructed by using the Query list in the data pane, just like tables. Facilities are provided for displaying tool tip information such as actual data values.

There are two sets of pop-up menus in Preview mode, depending on whether the cursor is positioned over the chart area or not. Right-clicking on the chart brings up a menu similar to that shown in Display D.8, displaying the various navigation actions available (applets only). Right-clicking outside the chart area shows a menu with choices for customizing the graph, including:

- Chart type – bar, pie, scatter, or segmented bar
- Visuals – color, legend, and 2D vs. 3D, depending on the chart type selected
- Variables – category and response
- Mouse Help – shortcut commands

A *Critical Success Factor* (CSF) component is a graphic display of a single value. Display D.10 shows the CSF for the sum of predicted sales:

Display D.10 Critical Success Factor

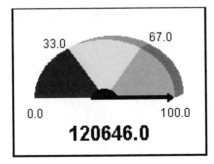

In order to specify an analysis variable, right-click on the image in Design view and select Properties from the pop-up menu. To choose a column and a statistic, select the Data tab from the resulting properties menu, shown in Display D.11:

Display D.11 Select Critical Success Factor Properties

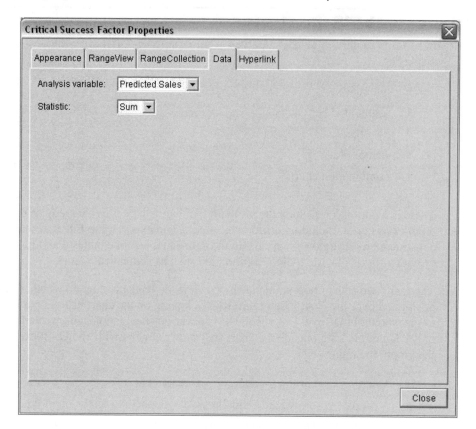

It is not immediately obvious, but the **Analysis variable** drop-down list in Display D.11 only includes measures that already appear in the document. If Sales Ratio for example, is not in the table, you cannot include it as a CSF.

Since it can sometimes be difficult to tell which subset is being viewed, the *drill path* component can be used to define labels for different subsets of the data. Drag a drill path control from the standard toolbar and drop it on the table. A box will appear that contains text strings for each of the subsets. Right-clicking the box and selecting the **LabelView** tab allows editing the labels; the **Appearance** tab is used to format the box.

In Display D.12, the font has been changed to bold and a simple rectangular box drawn around the text.

Display D.12 Drill Path Component

```
{Country: {Country}}
{Region: {Region}}
{Division: {Division}}
{Product type: {Product type}}
{Product: {Product}}
{Month: {Month}}
{Year: {Year}}
{Quarter: {Quarter}}
```

Switching to Preview displays the formatted labels for the data subset:

Display D.13 Drill Path Component Output

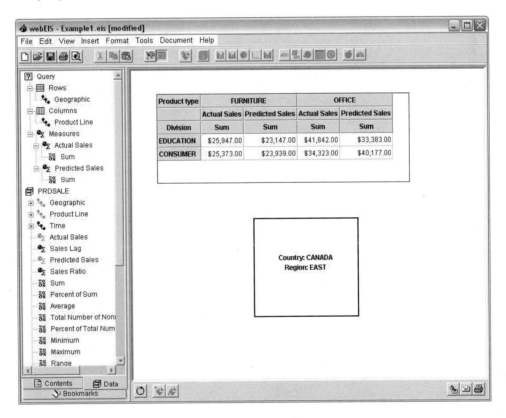

Finally, **Bookmarks** can be used to save the current state of a section. To create a public bookmark, available to all users, select the **Bookmarks** tab in the Navigation pane. Right-click **Public Bookmarks** and enter a name for the view. Users can now quickly go to a particular selection just by clicking on the bookmark. Private bookmarks, on the other hand, are set up by the user. User-defined bookmarks are stored on the server. The webEIS page designer must also specify a data set on the server where the bookmarks are to be stored; the details for doing this are available in the webEIS 2.0 online Help in the "Enabling Run-Time Bookmarks" topic.

Index

Books Available from SAS® Press

Advanced Log-Linear Models Using SAS®
by **Daniel Zelterman**

Analysis of Clinical Trials Using SAS®: A Practical Guide
by **Alex Dmitrienko, Geert Molenberghs, Walter Offen,** *and*
Christy Chuang-Stein

Annotate: Simply the Basics
by **Art Carpenter**

*Applied Multivariate Statistics with SAS® Software,
Second Edition*
by **Ravindra Khattree**
and **Dayanand N. Naik**

*Applied Statistics and the SAS® Programming Language,
Fifth Edition*
by **Ronald P. Cody**
and **Jeffrey K. Smith**

An Array of Challenges — Test Your SAS® Skills
by **Robert Virgile**

*Carpenter's Complete Guide to the SAS® Macro Language,
Second Edition*
by **Art Carpenter**

The Cartoon Guide to Statistics
by **Larry Gonick**
and **Woollcott Smith**

*Categorical Data Analysis Using the SAS® System,
Second Edition*
by **Maura E. Stokes, Charles S. Davis,**
and **Gary G. Koch**

Cody's Data Cleaning Techniques Using SAS® Software
by **Ron Cody**

*Common Statistical Methods for Clinical Research with
SAS® Examples, Second Edition*
by **Glenn A. Walker**

The Complete Guide to SAS® Indexes
by **Michael A. Raithel**

*Data Management and Reporting Made Easy with
SAS® Learning Edition 2.0*
by **Sunil K. Gupta**

*Debugging SAS® Programs: A Handbook of Tools and
Techniques*
by **Michele M. Burlew**

*Efficiency: Improving the Performance of Your SAS®
Applications*
by **Robert Virgile**

The Essential Guide to SAS® Dates and Times
by **Derek P. Morgan**

The Essential PROC SQL Handbook for SAS® Users
by **Katherine Prairie**

*Fixed Effects Regression Methods for Longitudinal Data
Using SAS®*
by **Paul D. Allison**

Genetic Analysis of Complex Traits Using SAS®
Edited by **Arnold M. Saxton**

A Handbook of Statistical Analyses Using SAS®, Second Edition
by **B.S. Everitt**
and **G. Der**

Health Care Data and SAS®
by **Marge Scerbo, Craig Dickstein,**
and **Alan Wilson**

The How-To Book for SAS/GRAPH® Software
by **Thomas Miron**

*In the Know ... SAS® Tips and Techniques From
Around the Globe*
by **Phil Mason**

Instant ODS: Style Templates for the Output Delivery System
by **Bernadette Johnson**

*Integrating Results through Meta-Analytic Review Using
SAS® Software*
by **Morgan C. Wang**
and **Brad J. Bushman**

Learning SAS® in the Computer Lab, Second Edition
by **Rebecca J. Elliott**

The Little SAS® Book: A Primer
by **Lora D. Delwiche**
and **Susan J. Slaughter**

The Little SAS® Book: A Primer, Second Edition
by **Lora D. Delwiche**
and **Susan J. Slaughter**
(updated to include SAS 7 features)

The Little SAS® Book: A Primer, Third Edition
by **Lora D. Delwiche**
and **Susan J. Slaughter**
(updated to include SAS 9.1 features)

The Little SAS® Book for Enterprise Guide® 3.0
by **Lora D. Delwiche**
and **Susan J. Slaughter**

support.sas.com/pubs

A Step-by-Step Approach to Using SAS® for Univariate and Multivariate Statistics, Second Edition
by **Norm O'Rourke, Larry Hatcher,**
and **Edward J. Stepanski**

Step-by-Step Basic Statistics Using SAS®: Student Guide
and *Exercises*
(books in this set also sold separately)
by **Larry Hatcher**

Survival Analysis Using SAS®:
A Practical Guide
by **Paul D. Allison**

Tuning SAS® Applications in the OS/390 and z/OS
Environments, Second Edition
by **Michael A. Raithel**

Univariate and Multivariate General Linear Models:
Theory and Applications Using SAS® Software
by **Neil H. Timm**
and **Tammy A. Mieczkowski**

Using SAS® in Financial Research
by **Ekkehart Boehmer, John Paul Broussard,**
and **Juha-Pekka Kallunki**

Using the SAS® Windowing Environment: A Quick Tutorial
by **Larry Hatcher**

Visualizing Categorical Data
by **Michael Friendly**

Web Development with SAS® by Example
by **Frederick Pratter**

Your Guide to Survey Research Using the SAS® System
by **Archer Gravely**

JMP® Books

JMP® for Basic Univariate and Multivariate Statistics: A Step-by-
Step Guide
by **Ann Lehman, Norm O'Rourke, Larry Hatcher,**
and **Edward J. Stepanski**

JMP® Start Statistics, Third Edition
by **John Sall, Ann Lehman,**
and **Lee Creighton**

Regression Using JMP®
by **Rudolf J. Freund, Ramon C. LIttell,**
and **Lee Creighton**